S0-CUG-412

Childhood Sexual Learning

Childhood Sexual Learning

The Unwritten Curriculum

Edited by
Elizabeth J. Roberts

Ballinger Publishing Company • Cambridge, Massachusetts
A Subsidiary of Harper & Row, Publishers, Inc.

This book is printed on recycled paper.

International Standard Book Number: 0-88410-374-9

Library of Congress Catalog Card Number: 79-27545

Printed in the United States of America

Library of Congress Cataloging in Publication Data

Main entry under title:

Childhood sexual learning.

Includes index.
1. Sex (Psychology)—Social aspects. 2. Nature and nurture.
3. Sex instruction for children. I. Roberts, Elizabeth J., 1944–
BF692.C54 155.4'3 79-27545
ISBN 0-88410-374-9

Contents

List of Figures and Tables

LIST OF FIGURES

LIST OF TABLES

Preface

This book is based on the premise that human sexuality and the ways in which we express it are not entirely inborn. Instead, maleness and femaleness are made up of a complex set of roles and attitudes that are developed and limited by family, society, and culture. Sexuality, moreover, does not emerge all at once—either at birth, at puberty, or in the marriage bed. It is substantially influenced by the learning experiences we have throughout our lives, and many factors have an impact on the development of sexual roles, relationships and life-styles.

In 1974, John D. Rockefeller 3rd established the Project on Human Sexual Development to (1) stimulate new thinking and develop new programs to expand understanding of human sexuality; (2) provide forums and agendas for discussion of the many complex personal and social issues that sexuality raises; and (3) consider the ways in which social and educational policies and priorities could improve the conditions of sexual learning in our society. During the past six years the Project on Human Sexual Development has explored various aspects of childhood sexual learning in our society. Particular emphasis has been given to the roles parents, community service providers, and television programs play in the sexual learning of young people. The more closely we examined these obvious and immediate influences on childhood development, the more apparent it became that they can only be adequately understood within the larger context of cultural values, social policies, and institutional traditions and practices.

All too frequently the relevance of institutional influences on sexual learning are ignored by social scientists, program planners, and policymakers. Most approaches to sexuality and sexual learning emphasize the individual (or the family unit); articles, books, advice columns, and service programs imply that it is up to each person to achieve sexual well-being through inner resources. Unfortunately, such a perspective fails to appreciate how powerful are social, economic, political, and cultural conditions in which sexual values are perceived and understood.

To focus attention on the broader social influences on sexual learning, the Project on Human Sexual Development invited the contributors to this volume to examine some of the more salient environmental influences on childhood sexuality. Authors were selected for their particular expertise in a field deemed important to our investigation—social welfare, work, religion, peer communication, urban design, telecommunications, education, and social policy. In their essays, the authors use their knowledge to expand our thinking about the impact that our social systems may have on different dimensions of sexuality.

It should be noted that each author's views are her or his own and not necessarily the views of the editor or other contributors. In many of the essays prepared for this book, there was little or no existing research that shed light directly on our concerns. Often the authors were required to speculate about the assumptions about sexuality that underlie institutional policies and organizational practices and the impact these assumptions may have on the development of sexual attitudes, values, and behavior patterns.

In an attempt to provide a conceptual framework for understanding human sexuality, Chapter 1 opens the inquiry into childhood sexual learning by suggesting the need for an expanded working definition of sexuality; a range of issues essential to a complete understanding of the process of sexual learning is explored. Chapter 2 provides the reader with an overview of social learning theory and the potential impact of key environmental influences on childhood development. Each of Chapters 3 through 9 focuses on one particular learning environment and attempts to assess its relevance to sexual learning in childhood. In a suggestive, if not exhaustive, manner these papers describe both the process by which messages about sexuality are communicated and the content of the messages. Finally, the last chapter synthesizes the environmental messages discussed throughout the volume and suggests the implications of this "unwritten sexual curriculum" for future policy planning.

Discussion of any complex topic requires some shared understanding. Unfortunately, in the area of sexuality and sexual learning there is little consensus about the scope of issues involved. This problem is further exacerbated by the lack of clarity in terminology. The term sexuality itself is imprecise, referring variously to a process of development, a dimension of personality, and an expression of behavior. The inconsistencies of terminology from chapter to chapter within this volume reflect not only the various perspectives of the authors, but also the current state of disagreement over terms. As a first step toward better understanding, the editor would consider "sexuality" to be the largest unit of analysis, within which such specific components as "gender role" and "erotic conduct" could be discussed. (Chapter 1 further elaborates a working definition of sexuality and defines the parameters of sexual learning's salient issues.)

This volume is the third in a three-part series on sexuality produced under the auspices of the Project on Human Sexual Development's Special Studies Program. Under the direction of Dr. Herant Katchadourian, Professor of Psychiatry & Behavioral Sciences, Vice Provost & Dean of Undergraduate Studies, Stanford University, this Program has attempted to raise issues, stimulate research, and increase understanding about sexuality and sexual learning among academicians, other professionals, and policymakers.

In the first volume, *Human Sexuality: A Comparative and Developmental Perspective** a distinguished group of biological and behavioral scientists review the most up-to-date research on sexuality from five perspectives: evolution, biology, psychology, sociology, and culture. Another volume in this series will discuss alternative theoretical perspectives for understanding the process of sexual learning from birth to adolescence. This volume extends the exploration of sexuality to an examination of the social contexts in which sexual learning occurs.

The diversity of environments discussed and the range of viewpoints represented in this volume do not comprise an integrative or exhaustive assessment of our society's sexual curriculum. Lacunae are an inevitable outcome of the complexity of the subject matter and a reflection of the current level of knowledge. It is our hope that the ideas presented here will stimulate discussion, debate, and further inquiry.

*Herant Katchadourian, ed., *Human Sexuality: A Comparative and Developmental Perspective* (Berkeley: University of California Press, 1979).

Acknowledgments

I would like to acknowledge my profound gratitude to Mr. John D. Rockefeller 3rd who six years ago initiated the Project on Human Sexual Development. He was a man who greatly affected my thinking and my life, and therefore is a very real contributor to this volume. As an individual, he was deeply and personally concerned about many of the issues raised in this book; as a philanthropist, he sought new ways to help others understand sexuality as a positive dimension of human life. In one meeting sponsored by the Project several years ago he said, "... human sexuality is a powerful force in the lives of all of us. Throughout our lives it manifests itself in many ways. It can be wonderful and beautiful or hurtful and negative.... We know our society is not doing as good a job as it should in helping us understand this important part of our humanity." I hope this volume and our continuing efforts may help others to understand and experience the richness and fullness of human sexuality.

I would also like to thank Joan Dunlop who—since the Project's inception—has supported and challenged me in ways that I hope have kept my thinking fresh and relevant. I appreciate the support for the Special Studies Program provided by the Carnegie Corporation of New York, and I am especially grateful to Barbara Finberg for her active involvement, encouragement, and assistance throughout the many changes and transitions of the Project's life. I thank John Gagnon who first persuaded me of the important influence of social learning on sexuality and with whom sharing ideas and philosophy has been a pleasure and a privilege. Herant Katchadourian deserves special thanks for his professional guidance and his dedication in seeing the series through to its completion. All of the chapters in this book benefited from the skilled writing talents and editorial judgments of Jonas Weisel; his contributions encouraged all of us to clarify our ideas as well as our syntax.

Finally, for this volume as with every endeavor undertaken by the Project, I am indebted to the support, enthusiasm, patience, and skills of Steven A. Holt, Ann Tagan Gillis, and Kathleen Bryar. The commitment and affection of the entire staff has made my work over the past six years a source of great joy and personal growth.

Chapter 1

Dimensions of Sexual Learning in Childhood

*Elizabeth J. Roberts**

In recent years, numerous organizations and countless professionals have stressed the centrality of sexuality to an individual's total personality.[1] Books and magazine articles have reminded us of the importance of sexuality in the establishment of personal identity, and in the formation of human relationships. However, despite this growing recognition of the far-reaching role that sexuality plays in our daily life, the evidence continues to mount that for millions of Americans sexuality is a source of confusion and anxiety.[2] Pervasive ignorance and misunderstanding about sexuality contribute to poor self-image, faulty communication, misinformed decisionmaking, and unnecessarily stressful personal relationships. Most distressing of all, the situation is self-perpetuating. A population that is confused or deals uneasily with sexual issues is unlikely to guide the sexual learning of its young people effectively and meaningfully. Today, many children, adolescents, and adults need and want help in understanding and appreciating the full dimensions of human sexuality.

Such understanding, however, does not come easily. Media attention usually focuses on sexuality as a source of personal or social

*Elizabeth J. Roberts is President of Population Education, Inc., and for the past 6 years served as Executive Director of the Project on Human Sexual Development, an enterprise committed to increasing the public's understanding of human sexuality. Ms. Roberts has directed a number of national policy inquiries including task forces for the White House Conferences on Children and Youth and the Children's Television Unit of the Federal Communications Commission. She has written numerous articles and is coauthor of *Family Life and Sexual Learning*, a major study of parent-child communication about sexuality.

problems, thus furthering the apprehension and anxiety of many individuals. Often, too, because public discussion is emotionally charged and cliché ridden, it adds more heat than enlightenment to the consideration of needed educational and social initiatives. Foremost among the barriers, two widely held misconceptions serve as obstacles to a full understanding of the issues: (1) the belief that sexuality is limited to sex and reproduction, and (2) the assumption that sexual learning is limited to formal, compartmentalized programs.

SEXUALITY IS MORE THAN SEX

For many individuals the word sexuality brings to mind a brief and limited encounter that begins and ends at a certain time, occurs with a certain person in a certain place, involves specific parts of the body, and frequently results in conception. This narrow view of sexuality as something separate from the rest of life leads many people to ignore other important channels through which sexuality is experienced and expressed. It often creates a preoccupation with sexual "performance" and "adequacy," which in turn results in anxiety, misunderstanding, and dysfunction. The same narrow view fosters a perspective concerned essentially with women and reproduction. Many educators and service providers, therefore, neglect the important role sexuality plays in the lives of men and the vital role that men play in sexual decisionmaking.[3] Finally, the emphasis on the erotic and reproductive dimensions of sexuality means that during childhood the discussion of sexuality is strongly subjected to ambivalence and taboo. Believing the age to be one of "sexual innocence," many parents and professionals are predisposed to ignore the importance of childhood sexual learning.

A new, more expansive definition states:

> Human sexuality is not confined to the bedroom, the night time, or to any single area of the body. It involves what we do, but it is also what we are. It is an identification, an activity, a biological and emotional process, an outlook and an expression of the self. . . . It is an important factor in every personal relationship and in every human endeavor from business to politics.[4]

Understood in this larger context, sexuality involves considerably more than intercourse, reproduction, or social hygiene. Sexuality is part of our basic identity. It encompasses our total sense of self. It involves our attitudes, values, feelings, and beliefs about masculinity and femininity; it contributes to our feelings about our physical selves—the limits, joys, and embarrassments of our bodies. It is the integration of our needs for affiliation and intimacy, and our expressions of love and affection, as well as our fears, fantasies, and decisions regarding our erotic conduct. Human sexuality is expressed in all of our interactions with others. It influences and is influenced by our interpersonal relationships, our family roles, and our social lifestyles. "Sexuality is emotional, physical, cognitive, value-laden, and spiritual. Its dimensions are both personal and social. . . . Sexuality is a central dimension of each person's selfhood, . . . a critical component of each person's self-understanding and of how each relates to the world."[5]

SEXUAL LEARNING VERSUS SEX EDUCATION

If we accept this broad-based meaning of sexuality in our lives, then education in sexuality includes much more than one parent-child talk about the "facts of life," or a sixth-grade course in anatomy and physical development, or a lecture on contraception and family planning. Sexuality is not an immutable, predetermined entity that proceeds uncontrollably along some predestined course, but is a malleable set of behaviors, fantasies, beliefs, and attitudes that can be changed and shaped by learning, choice, and action.[6] Sexual learning is social learning. It begins at birth and many sexual attitudes, beliefs, and behavior patterns are influenced by childhood experience and learning. It is important to remember that individuals do not become sexual all at once—at birth, at puberty, or in the marriage bed. Identity grows slowly; it builds on the many interactions that define who each growing person is and how each fits into our social structures. Childhood learning is an important—but not totally determining—period in this process.

The first years of a child's sexual learning are integrally related to the development of social, mental, emotional, and physical capacities. During infancy and early childhood, children become aware of their bodies—that is, they become aware of the body's shape, size, functions, and capacities for pleasure. They develop a sense of psychological self; they begin to learn socially approved gender roles; and they develop some understanding about acceptable behavior and close relationships. They begin to develop attitudes and behavior patterns that can influence the way in which they internally experience and externally express their sexuality throughout life.

Although it has often been assumed that the major component of sexual learning is direct verbal instruction, in recent decades we have learned that much human communication is carried on nonverbally through body postures, facial expressions, and other behavior patterns. This is true of learning about sexuality as well; children learn in large measure without direct instruction, but through observation and imitation. Think of the process by which children learn to speak the language of their family or their cultural group. With little direct instruction, their speech patterns mimic the accents and intonations of significant others in their lives. Extending this speech analogy further, we might ask whether formal sex education in the school is any different from formal instruction in Latin. Although most students learn some Latin in classes and may become more self-aware about their language use, the Latin language does not replace the family tongue in students' daily discourse.[7]

An emphasis on nonverbal communication is not intended to dismiss the important role of direct verbal communication in a child's sexual learning. If a young child receives an inaccurate response to a specific question, he or she may go through life making decisions based on erroneous information. Or the child may discover the correct answer on his or her own and grow to distrust the parent, teacher, physician, or minister who offered the first answer. If the naturally inquisitive child is rebuked, ridiculed, or ignored when seeking information about some aspect of sexuality, he or she may learn to stop asking questions of authority figures and may rely more heavily on (frequently inaccurate) peer information. Worse, the child may grow up unable to talk about sexual feelings and needs.

Many marriage counselors and sex therapists have observed that couples often cannot communicate about the simplest of their sexual interactions and that many people are ignorant of the basic sexual needs or erotic responses of their partners. If fears or inhibitions regarding verbal communication about sexual matters are perpetuated, adolescents and adults may have difficulty in communicating the meaning of sexual events in their lives, or may misunderstand the sexual psychology of another person. Both verbal and nonverbal communication are important to the process of sexual learning in childhood.

Children learn about sexuality from their everyday experiences, through their observations and interactions with parents, peers, neighbors, and relatives. They learn from television programs, books, and toys. They also develop attitudes and values from the subtle—and sometimes not so subtle—messages about sexuality that are communicated through the very structure of our social institutions. Work policies, day-care facilities, religious services, doctors' offices, or welfare programs can contribute to or limit a growing child's sexual learning.

As Mary Jo Bane and Steven A. Holt note in Chapter 7 below, while many medical schools now provide some training in sexual development for obstetricians, gynecologists, or psychiatrists,[8] very few schools offer such training for pediatricians. Perhaps behind this disparity lies the implicit assumption that pediatricians, who deal primarily with infants and children, will have no opportunities to use information about sexuality. In any case, the result is that it is difficult for the pediatrician, the parents, or the child to ask questions about masturbation, sex play, or other aspects of childhood sexuality.

Another example of the potential influence of social structure on sexual learning can be found in the world of work. Many work policies and practices are based on outmoded assumptions about the structure of family life. One such premise has been that if a woman—particularly if she were also a wife and mother—could afford to stay home and not work at a paying job, she would do so. As a result, few workplaces today are responsive to the needs of the growing segment of the labor force made up of women with children, or to the needs of the husbands and families of these women. Conventional work schedules often conflict with school-opening times. Lack of maternity leave or leave to care for an ill child frequently creates financial and emotional stress for families dependent upon the income of two wage earners.[9] Men seeking paternity leave or requesting time to be with their children are viewed as social anomalies. The growing child is not impervious to the demands and limitations that these structures impose on his or her parents and the organization of family life.

Such underlying assumptions about sexuality, unexamined for generations, form part of the social curriculum that guides the development of sexual roles, relationships, behavior, and life-style. Certainly, the child's experience of this curriculum changes as the child grows: experiences and interactions broaden; needs and interests shift; and capacities for organizing experiences and making sense of patterns expand. Thus, what is specifically learned about sexuality, how it is learned, and who the "teachers" are may be different at different stages of the life cycle. However, the learning process goes on all the time, everywhere. We are both teachers and pupils in this process, from birth until death.

DIMENSIONS OF SEXUALITY

One of the most difficult barriers to any consideration of the process of sexual learning is the lack of consensus about the range of issues involved. For many people, learning about sexuality means receiving facts about physical development and human reproduction; to others, sexual learning is the acquisition of knowledge about erotic behavior, or the development of a social life-style; and for many, it is simply an unnecessary consideration of an innate and biologically determined drive.

This section postulates a framework for understanding various aspects of human sexuality. These aspects are: gender role, affection and intimacy, family roles and social life-styles, body image, erotic experiences, and reproduction. Since comprehensive consideration of each area is beyond the scope of this chapter, the approach has been necessarily selective and impressionistic. Nonetheless, these six content areas together suggest a dynamic definition of sexual learning that unifies various dimensions of sexuality. The list is not exhaustive. It focuses principally on those aspects of sexuality that are socially learned. (This approach is not intended to deny the influence of biology; the effects of anatomy, chromosomes, and hormones on our sexual development are widely recognized. However, to say that our sexuality has a biological substrate is not to say that it is biologically determined.) This perspective chooses to emphasize the degree to which human beings are socially adaptive rather than chemically driven organisms. As social beings, we live among people within social, cultural, and political structures. Our interactions with these people and systems have an important influence on how our sexual attitudes, values, and behavior develop.

It should be pointed out that while each area is addressed separately in the following paragraphs, each aspect of sexual learning is essentially interwoven with the others—both theoretically and in the fabric of everyday life. Changes in one area are likely to bring about changes in the others. Given this interaction, the areas need to be viewed as dimensions of the same topic.

Gender Role

The single most important area of learning affecting our sexuality is the set of messages that we receive throughout our lives about "appropriate" masculine and feminine roles. Our cultural expectations have been and still remain gender-specific. It is difficult to escape gender-specific influences on our identity as sexual beings. From birth we are socialized according to our gender to exhibit particular skills, traits, and mannerisms; to express or control emotions; and to aspire to certain life goals.

Not long ago, the question of how a person develops an ongoing sense of self as either masculine or feminine was thought to be nearly irrelevant. Boys became men and demonstrated masculine traits and behavior; girls became women and behaved in feminine ways. It was assumed that there was a linear and necessary relationship between biological sex and specific gender roles.

Considerable research in recent years, however, has demonstrated the strong impact that social learning has on the development of gender traits and behavior.[10] Numerous studies have clarified the ways in which boys and girls, as well as men and women, are segregated from each other in our society—both physically and psychologically. Physically, boys and girls are encouraged to play in different ways with different toys, to participate in different elective school curricula, and to expect to assume different responsibilities within the family, community, and society. Even more important, society also segregates males and females psychologically by reinforcing the notion that if one gender behaves one way, the other should not. If girls can cry, boys should not. If women are nurturant, men are not. If females are vulnerable, males are tough. Even the phrase "opposite sex" implies this gender segregation.

These fundamental gender-role assumptions, unchallenged for decades, are now being scrutinized. Many individuals have begun to question the meaning and utility of traditionally segregated definitions of masculinity and femininity. And for many, it is becoming difficult to think in dichotomous masculine and feminine terms.

In many academic and professional circles, the notion of androgyny is increasingly used.[11] Androgyny is the psychological state in which an individual incorporates traits that are associated with both traditionally masculine and feminine attributes. Thus, an androgynous person is one who—depending on circumstances and regardless of biological sex—can be either independent and assertive or dependent, nurturant, and tender. With an increase in this androgynous gender-role perspective, other aspects of sexual learning are likely to become more complex and—in the minds of many persons—more humane and satisfying.

For many young people and their families, the relaxation of rigidly defined standards and norms for masculine and feminine roles has meant increased freedom and expanded opportunities for self-exploration and fulfillment. However, for many others, the nagging inconsistencies between traditional gender values and the challenges of new social realities have caused confusion and increased uncertainty about the meaning of sexual roles, relationships, and behavior patterns in our society. Whatever the individual perspective, the deliberate and unconscious gender training that occurs throughout childhood is a major influence on the sexual values and decisions of our youth.

Affection, Love, and Intimacy

As social beings we communicate many messages about our needs for affection and affiliation with others. Through our words and actions we convey how, when, and with whom we consider it appropriate or desirable to share our intimate thoughts and selves. From these experiences the growing child learns acceptable ways of interacting intimately and affectionately with others. The way in which young people learn to develop friendships, to gauge a relationship's depth, to reveal confidences and frailties, powerfully affects their sexual self-concept.

For most people, the development of attitudes about love and affection, the capacity for closeness, the ability to ask for and give comfort and support are significantly influenced by their gender training. In our culture, males are often encouraged to be less expressive and sensitive, and more emotionally distant than females. At the same time, females are encouraged to be openly affectionate, solicitous, nurturant, and sensitive.

Studies of childhood friendships reveal that girls, in their intensive relationships with one or two best friends, have more opportunities to develop "empathy skills"—to practice being intimate and to disclose their thoughts, feelings, and experiences to friends—than do boys.[12] Boys, on the other hand, tend to play in larger groups of same-sex peers, and find it difficult to learn ways of expressing their feelings of vulnerability or affection with one another. Feelings of anxiety on the part of parents, teachers, friends, and relatives regarding "male sexuality" may contribute to the discouragement of close intimate friendships between two boys. Such intimate dyadic friendships between boys may trigger fears of either "unmanly behavior" or potential homosexuality. As young people grow up social norms, work expectations, and family structures further perpetuate these differences in interpersonal relationships. Only recently have therapists and counselors drawn attention to the price each gender has paid for these divisions in the experience and expression of intimacy.[13]

Human sexuality encompasses the ability to participate with others in relationships at varying levels of intimacy. The capacity to reveal intimate thoughts and feelings to another person and, in turn, to receive someone else's self-disclosures is an important part of sexual learning.

Family Roles and Social Life-Styles

The pattern and meaning of our close relationships and our expressions of intimacy are often communicated through the ways in which we arrange, integrate, and manage our life-styles.[13] Based on personal and cultural views about appropriate sexual and social roles and relationships, young people learn to structure their contact with members of the same and the other sex. Adult decisions to get married, to get divorced, to remain single, to live alone or with friends, to balance work and family responsibilities, all contribute to the child's understanding of acceptable interpersonal interactions, family patterns, and life-styles.

Contrary to the clearly articulated expectations of only a few decades ago—when marriage, parenthood, and Mom-stay-at-home-while-Dad-works was the accepted adult norm—children today are confronted with a much wider array of social life-styles. It is virtually impossible for anyone who reads a daily newspaper, switches on a situation comedy, peers into the executive lunchroom, or visits a day-care center at 5 P.M. to deny the fact that there have been substantial changes in the way Americans live. The working wife and mother, the single parent, the sexually active teenager, the unmarried, cohabiting couple are no longer social anomalies. They are our friends, our siblings, our parents, our colleagues, our children, ourselves.[14]

Changes occurring in social life-styles and family roles raise some of the most complex issues individuals face today. As in many other areas, this freedom of choice is a mixed blessing. Many people—both men and women—who are living alone, in partnerships, or in families (of various shapes and sizes) are experiencing difficulties as traditional expectations make way for new life possibilities and demands. On the one hand, there is increased legitimacy in choosing any one of a number of life-styles (a word foreign to our grandparents as they were growing up). On the other hand, the taken-for-granted life path has been replaced by a myriad of new decisions that involve clarifying values, assessing personal needs and desires, and negotiating changes with partners or friends.

The social situation requires young people to develop decision-making skills and the ability to work out systems for directing themselves and controlling external events, and to find satisfying and responsible ways to structure their sexual and social contexts. Such skills require self-awareness, as well as understanding about sexuality and its impact on life.

Body Image

Many experiences throughout life contribute to the development of body image, and these feelings and beliefs about our bodies are inextricably linked to sexual identity. Early experiences of cuddling, touching, hugging, and holding teach children that their bodies are a source of pleasure and self-knowledge. So important are these activities to total well-being that if such sensory stimulation is withdrawn, children will often fail to thrive.[15]

Whether positive or negative feelings about one's body are fostered as one grows older is largely a matter of social learning. For example, in families (or in day-care centers, doctors' offices, schools, etc.) in which touching the genitals is followed by an admonition or slap, in which questions about physical development are greeted with anxious silence, or in which masturbation is deterred by the threat of physical or mental harm, young people may learn that understanding their bodies or experiencing them as pleasurable is somehow wrong. Similarly, cultural attitudes toward nudity, modesty, and body contact also influence whether the body is viewed as a source of pride and delight or a source of shame and embarrassment.[16]

Unfortunately, in today's society most boys and girls are not given opportunities to know or understand their bodies. They have been taught that asking questions about the body is inappropriate and that exploring or touching body parts, especially the genitals, is harmful. These learning experiences can make simple health precautions in later life—such as self-examination of the breasts, a check-up for venereal disease, or use of a condom or a diaphram—unnecessarily fraught with anxiety. Furthermore, serious difficulties in communicating bodily needs and feelings can develop, and sexual decisions may be made on the basis of noninformation or misinformation.

As in the other areas of sexual learning in our society, gender role and body image are closely allied. Traditionally, the emphasis for males is on performance and skill development—"What can my body do?" Boys' games that are based on power, aggression, contact, speed, and skill serve to socialize young males into seeing their bodies as effective instruments over which they exert control, and with which they must prove their masculinity. On the other hand, the emphasis for females is on attractiveness and desirability—"How do I look?" Girls learn to view their bodies as instruments of attraction, and frequently come to need the approval of others to validate their femininity and their physical desirability. Liking our bodies is part of liking ourselves; what we come to believe about how our bodies should feel, how they should look, and how they should work will exert an important influence on how we perceive and express our sexual selves.

Erotic Experience

For most children, learning about the erotic dimensions of sexuality is a covert and slightly illicit experience. Childhood eroticism is denied, and few children ever have a parent (or other adult) actually discuss erotic feelings and behavior with them. Nevertheless, through peer communication and through incidental observation, most boys and girls in our society learn to perceive and interpret various activities as erotic or "sexy," and to equate these with genital intercourse.

In our society interest in human sexuality centers around sex—specifically sexual intercourse. In part, this preoccupation with genital functioning has been encouraged by the (often pioneering) work of sex researchers and therapists—or, to be perhaps more accurate, by the mass media coverage of their work. From Kinsey to Masters and Johnson to Shere Hite, widely publicized books and articles focus public attention on the predominantly physical aspects of genital behavior. To look at most of the popular literature, one would think that genital behavior constitutes the bulk of human sexuality and that sexual awareness is a matter of understanding the most popular techniques aimed at the attainment of the ultimate orgasm.

This emphasis on intercourse results in a very narrow frame of reference for understanding human sexuality in general. In addition, it can limit understanding of other aspects of erotic experience as well. Instead of learning to view sexual expression as a central component of daily life, we are taught to identify sexuality as a brief and limited encounter that is assigned to its proper place (the bedroom) and time (following the eleven o'clock news).

Equating sexuality with intercourse virtually severs it from the context of life itself, and the consequences for sexual understanding can be unfortunate. For example, many young people (and adults) learn to deny or to feel guilty about their need for other forms of sensual contact. Being touched, stroked, or held—that is, having one's "skin hunger" satisfied—is a deep sexual need for most people. Likewise, preoccupation with genital functioning can cause individuals to neglect fantasy, memory, and such sensory aspects of eroticism as soft lights, romantic music, and enticing aromas. These "intangibles" often are what make erotic experiences full, rich, and satisfying. Finally, when other aspects of eroticism are ignored, the approach to sexual activity may become excessively goal-oriented and preoccupied with performance—the result of which is the current cult of orgasm. Thus, if one learns to identify sexuality totally with genital performance, dysfunction in this area may call into question one's entire sense of sexual self-esteem. While orgasm may be *nonpareil*, it is not the *sine qua non* of human sexuality.[17]

Reproduction

It is a fact that certain sexual activity can and does result in new life. A discussion of sexual learning would be incomplete without consideration of this aspect of human sexual functioning. Most of us want our children to live personally satisfying and socially reponsible sexual lives. But enjoying oneself sexually involves decisionmaking and management; freedom involves control over our lives. And most young people seem to learn about sexuality as if it were something over which they had little or no control. It is difficult to lead an anxiety-free sexual life if one is uninformed or misinformed about pregnancy, contraception, venereal disease, family planning, or abortion.

However, while learning factual information about these aspects of reproductive sexuality is important, it is insufficient and incomplete outside of the cultural, social, and psychological context of human sexuality. Even if children know the "facts of life," they need opportunities to clarify their values, assess the consequences of their desires, and develop decisionmaking skills.

Another important aspect of reproduction in sexual learning is the history that reproduction has had in guiding our cultural perspective on human sexual conduct in general. Historically, many people have learned that the only "good sex" is sex that could possibly result in a socially sanctioned conception and pregnancy. Accordingly, all other forms of sexual expression are bad, and sexual intercourse is only a reproductive act. Engaging in sexual activities for "fun" is not deemed a responsible expression of human sexuality. One result of this historical perspective has been to generate two groups of people, the sexually acceptable and the sexually unacceptable—one the sexually elite and the other the sexually oppressed. The sexually elite include those who are married, heterosexual, healthy, young (but not too young), attractive, and rich (able to afford a child). The sexually oppressed include those who are single, "too young," "too old," poor, homosexual, physically disabled. mentally retarded, ugly, and so on.[18]

Clearly, looking at sexuality only through the perspective of reproduction produces a narrow and distorted understanding of sexuality. Yet to talk about sexual learning outside the context of reproduction is equally narrow. For many people, the most important sexual decision that they will make is the decision to have or not to have a child. And many sexual activities are shaped and colored by the experience of avoiding or welcoming conception. The concern for a planned and manageable family life, awareness of alternative contraception methods, and even concern for the larger issues of population have a fundamental place in a child's sexual learning.

CONCLUSION

As noted earlier in this section, the dimensions of sexuality discussed here are interrelated. Learning in one area can influence feelings and attitudes about other aspects of sexuality. The following example might serve to highlight this process.

Today, many young adolescents are faced with deciding whether or not to engage in sexual intercourse. Imagine two sixteen-year-olds (male and female) involved in this decision. Many of the messages that they have received about various dimensions of their sexuality will play a part in influencing their decision. Perhaps the young male was raised to learn—as part of his gender role—that growing up means to grow away from others, to be independent, and to know answers without having to ask questions. Perhaps he has had little training in expressing his feelings and his fears. He may have learned that his body is something he must perform with or use to prove himself and his masculinity. Perhaps he has been encouraged by peers, teachers, and parents to be goal-oriented and to succeed at all costs—even in the bedroom. Taught to identify masculinity with being cool or "scoring," taught that wanting to be touched or held or comforted might be perceived as weak or unmanly, our adolescent boy might be expected to decide that the best way he can express his masculinity or receive affection is by pressuring for intercourse.

Now for our young girl. Perhaps she was raised to believe in the importance of meeting the needs of others, of being nurturant and emotionally attendant. Perhaps billboards, TV programs, parents, and friends have conveyed to her that the most important aspect of her body is its attractiveness to males. Perhaps she was raised to think that "sex" is something that happens to her, something she knows very little about, and something about which she should not or cannot make decisions. Perhaps she is ambivalent or frightened about her own erotic feelings, and the only information she has about male physiology is the inaccurate notion that males are possessed by an uncontrollable sex drive.

Will the conditions of this sexual learning permit these adolescents to make an informed, satisfying, and responsible decision? Or will she have to pretend it all "just sort of happened"? ("I was swept off my feet.") Does he have to fear being thought less of a man if he really wants simply to be held? Do they have to worry "What will my parents (or teacher or doctor or minister) think if I ask them about contraceptives?" Have the ways in which they have learned about sexuality enabled them to talk openly and honestly with each other, to ask for guidance, or to trust advice?

This example is *not* about adolescent intercourse. Rather it is about the many messages—the sexual learning—that influence what we do and how we feel about what we do. It is about body image, about feelings of love and intimacy, about what we think is "sexy," and how we learn to make decisions.

The dimensions of sexuality highlighted in this chapter can only begin to suggest the range of human experience that influences sexual learning. This process of learning is not organized like a textbook or a lesson plan where first we learn this and then we learn that. Rather, it is often a chaotic and disorderly collection of learning that never quite becomes completely integrated. Most people are, or at least feel, alone when they are forced to make sense of their sexuality and the frequently conflicting sexual messages received throughout life. Children, adolescents, and adults are required to make their way to responsible sexual satisfaction without ever talking about responsibility, sexuality, or satisfaction. If these indeed are the conditions of learning about sexuality, then the evidence that many people find their own sexuality a source of difficulty should come as no surprise.

Because as a society we have not looked at the processes through which people become sexual and the role that sexuality plays in all our lives, we remain fearful that if we do anything, something will go wrong. The net result is another generation of individuals who are confused or misinformed about sexuality. Discussion about sexuality and sexual learning must be brought to the attention of the public in a way that will dispel the fear, anxiety, and misinformation that permeates most attitudes. Sexuality needs to be legitimized, and its effect on the quality of life of the whole community needs to be recognized.[19] The full range of topics must be explored in a form and in forums that can facilitate thoughtful new consideration.

This is no small task—putting a new idea on our nation's agenda. It amounts to nothing less than having us think differently about ourselves and the world in which we live. Yet that awareness is the first step in improving the conditions of sexual learning and increasing understanding about human sexuality.

NOTES TO CHAPTER 1

1. See statements by the American Medical Association, the World Health Organization, the United Church of Christ, the Planned Parenthood Federation of America, and the Sex Information and Education Council of the United States.

2. Commission on Population Growth and the American Future, *Population Growth and the American Future* (New York: New American Library, 1972).

3. U.S. Congress, House, Select Committee on Population, *Fertility and Contraception in the United States* (Washington, D.C.: Government Printing Office, 1978), pp. 2, 80–81.

4. American Medical Association Committee on Human Sexuality, *Human Sexuality* (American Medical Association, 1972), p. 3.

5. United Church of Christ, *Human Sexuality: A Preliminary Study* (New York: United Church Press, 1977), pp. 12–13.

6. Eleanor Morrison and Vera Borosage, eds., *Human Sexuality: Contemporary Perspectives* (Palo Alto: Mayfield Publishing Co., 1977).

7. Alberta Siegel, "Long-Term Outcomes of Early Childhood Experiences" (paper prepared for the Project on Human Sexual Development, 1975).

8. Harold Lief, *Medical Aspects of Human Sexuality*, 13, no. 2 (1979), p. 44.

9. Sheila Kammerman, "Parenting in an Unresponsive Society," in *Notes: Program in Sex Roles and Social Change* (New York: Columbia University Center for the Social Sciences, 1979).

10. Eleanor Maccoby and Carol Jacklin, *The Psychology of Sex Differences* (Stanford: Stanford University Press, 1974); A. Bandura, "Social Learning Theory and Identifactory Process," in *Handbook of Socialization Theory and Research*, ed. D.A. Goslin (Chicago: Rand McNally, 1969); Patrick Lee, and Robert Sussman Steward, eds., *Sex Differences: Cultural and Developmental Dimensions* (New York: Urizen Books, 1976).

11. See Sandra Bem, "Sex Role Adaptability: One Consequence of Psychological Androgyny," *Journal of Personality and Social Psychology*, 31, no. 4 (1975): 634–643; June Singer, *Androgyny: Toward a New Theory of Sexuality* (Garden City, N.Y.: Doubleday, Anchor Books, 1976); Carolyn Heilbrun, "Recognizing the Androgynous Human," in *The Future of Sexual Relations*, eds., Robert and Anna Francoeur (Englewood Cliffs, N.J.: Prentice-Hall, 1974).

12. Janet Lever, "Sex Differences in the Complexity of Children's Play," *American Sociological Review*, 43 (1978): 471–482.

13. See John Gagnon and Cathy Greenblat, *Life Designs: Individuals, Marriages, and Families* (Glenview, Ill.: Scott, Foresman and Co., 1978).

14. Judith Simpson, Lucille Aptekar-Litton, and Elizabeth J. Roberts, *Harmonizing Sexual Conventions: Service Providers and Sexual Learning* (Cleveland: Cleveland Program for Sexual Learning, 1978).

15. Siegel, "Long-Term Outcomes of Early Childhood Experiences."

16. "Children of Four Already See Genitals as Taboo Areas," *Sexuality Today*, 2, no. 50 (1979), p. 1.

17. F. Gerald Brown, "Human Sexuality Defined" (paper for L.P. Cookingham Institute of Public Affairs, Kansas City, Missouri, 1979), p. 40.

18. Ibid., p. 49.

19. United Church of Christ, *Human Sexuality*.

 *Chapter 2*

Toward an Understanding of Sexual Learning and Communication: An Examination of Social Learning Theory and Nonschool Learning Environments

*Janet Kahn and David Kline**

INTRODUCTION

The essential purpose of this chapter, indeed of this entire volume, is to encourage the adoption of a more comprehensive and purposeful approach to sexual education in the United States. A critical aspect of this process has been addressed in Chapter 1, that is, broadening and clarifying our understanding of those aspects of life that are a part of our sexual development or conduct.

The next step is to remove some of the mystery from sexuality, or at least from the sexual learning process. This mystery stems from popular explanations of sexual development that hinder us, both as individuals and as a society, from assuming responsibility for our own sexual conduct and values. One of these explanations attributes sexual behavior to a variety of intangible forces that guide us, but can barely be guided by us. Whether these forces are labeled human nature, hormones, or the devil can be significant, yet by the time they filter down to an individual person, family, or town, the basic message is that sexuality is largely beyond our control. Another

*Janet Kahn is a staff member at the Somerville (Massachusetts) Women's Center. As a research scientist at the American Institute of Research, she has worked in the area of teenage sexuality. She is executive producer of the videotape "Growing Up Fast: What Service-Providers Should Know about Teenage Parenthood" and coauthor of the research report "The Ecology of Help-Seeking Behavior among Adolescent Parents."

David Kline is Research Associate at the Harvard Graduate School of Education, Lecturer at the Harvard School of Public Health and Visiting Professor of Education, Boquazici University, Istanbul, Turkey. He has written and cowritten several books, including *Issues in Population Education*.

explanation places primary responsibility for psychosexual development on the guardian (usually the mother) of the young child. Although this explanation brings the possibility of a purposeful approach to learning more within human grasp, it really places a disproportionate responsibility on one group of people while allowing policy planners to ignore their responsibilities. It also discourages us unnecessarily from attempting changes later in life. Certainly there are more sophisticated versions of these explanations. However, until the whole subject receives more serious attention at every level, understanding at the popular level will not be likely to surpass these explanations.

In contrast, our position is essentially that put forth by John Gagnon in his book *Human Sexualities:*

> In any given society, at any given moment in its history, people become sexual in the same way they become everything else. Without much reflection they pick up directions from their social environment. They acquire and assemble meanings, skills and values from the people around them. Their critical choices are often made by going along and drifting. People learn when they are quite young a few of the things they are expected to be, and continue slowly to accumulate a belief in who they are and ought to be throughout the rest of childhood, adolescence and adulthood. Sexual conduct is learned in the same ways and through the same processes; it is acquired and assembled in human interaction, judged and performed in specific cultural and historical worlds.[1]

Thus our focus is on environmental influences in sexual development; that is, the learned aspects of such development. We emphasize these influences not to disallow the useful contributions of biological or developmentalist approaches but rather as an antidote to an overdose of those approaches and to interpretations of them that tell us that either no one, or mothers alone are responsible for the current understanding and expressions of human sexuality. We adopt a social learning perspective for balance, and suggest that society is responsible for sexual learning and that policy interventions, in all the realms indicated in Chapter 1 and throughout this volume, can make a difference.

This chapter therefore, presents a summary of those aspects of social learning theory that we believe to be particularly useful in guiding policy. At this stage a critical part of that work is the broadening of our understanding of the arenas of daily life from which we receive sexual messages. We make no pretense about delivering the definitive statement on these matters. Rather, we hope that others will enter the debate, bringing new research and concepts. Sexuality, particularly in this broadened definition, is at the core of our humanity, and we can no longer afford an ignorant or haphazard approach to our lives.

In this chapter, the term "social learning" is used to refer to the process by which a person or group acquires information, attitudes, values, beliefs, and behavior skills and patterns as part of the process of adjusting to and interacting with a social environment.[2]

The key word here is "interaction." The urge to establish social ties with other human beings may or may not be a universal instinct. Sexuality, too, may or may not be an instinct or biological drive. Our focus here is on the innumerable ways in which people establish patterns to meet their need for human relatedness. There are many ways in which a person experiences him or herself as, and behaves as, a sexual being. These variations in meaning and behavior are established in interaction, both with other people and with the structural environment.

The term "structural environment" includes social institutions as the family and the school, as well as the medical, legal, and economic systems. It can also include the subtle and particular aspects of social structure—the ratio of grass to asphalt in one's neighborhood, the seating arrangement in a classroom or office, the amount of time a family spends at the dinner table, the design of the building where one lives.

Furthermore, when we speak of an individual interacting with a social environment, we will be emphasizing conceptions of social learning that are "dynamic" and in which the effects flow in more than one direction. That is, people do not simply adjust to social institutions; people and institutions influence each other. The family is not simply a group in which parents care for children and teach them the ways of the world. It is a social system in which all members affect and learn from one another. One's perceptions influence experience; similarly one's experiences then influence one's perceptions of the self in the world.

Another area of assumptions and definitions requiring discussion is that of learning environments. It has been helpful for us to distinguish between "formal," "nonformal," "informal," and "incidental" learning environments and processes (see Figure 2–1). We distinguish among these four categories by the presence or absence of a structured curriculum and a certification system, and on the basis of whether the learner, the "teacher," or both are intentionally engaged in the educational experience. Formal education environments are schools. These are highly structured environments with relatively clear curricula and recognized processes of certification. Further-

Figure 2–1. Four Learning Environments.

Type of Learning	*Recognized Certification System*	*Intent to Teach*	*Intent to Learn*
Formal	√	√	√
Nonformal	0	√	√
Informal	0	√/0	0/√
Incidental	0	0	0

more, both learners and teachers are officially present for educational purposes.

Nonformal education refers to structured environments outside the school—such as Boy Scouts and Girl Scouts, adult education classes, and job training. In such situations there is some structure and curriculum, and all parties are present for educational purposes. However, if there is certification, it is usually not recognized as possessing the high degree of general value that formal schooling has.

Informal learning is less structured and occurs in situations with no acknowledged curriculum and no certification system. In informal learning only one party, either the learner or the teacher, has educational intentions. Examples would include a newspaper reporter learning on the job, striving to become an editor or a child learning how to behave from a parent's rewards and punishments.

Incidental learning refers to learning as a by-product of some activity, such as learning about family relations from the allocation of space in the home. There is no intent on the part of either the teacher or learner, and no structured curriculum or certification process.

This chapter will concentrate on nonformal, informal, and incidental learning environments and processes—areas that in the past have tended to be the most ignored aspects of social learning. We will look at formal school environments only insofar as they are the site of peer interaction.

Our exploration of social learning will begin with a discussion of the theoretical underpinnings of social learning theory. In particular, we will examine the concepts of conditioning, reinforcement, and modeling as they have been established through laboratory research. Next we will discuss both interpersonal and institutional-structural processes of communication. Then we will look at verbal and nonverbal, and intentional and unintentional aspects of human communication.

One of our theses is that a major source of messages delivered and perceived by the individual is the way in which other individuals and institutions allocate such key resources as time, space, energy, and money. From that theoretical base we will move to an exploration of the learning environments in everyday life. We will present examples of social learning processes as they occur in the home, at work, in groups, through the media, in interaction with professionals, and in religion. These sections correspond closely, but not exactly to chapters 3 through 10 of this volume. The sections in this chapter seek to identify and describe within each environment both the sources of sexual messages and the dynamics of the learning process. The later chapters add more detail to our understanding of specific dynamics, and go on to discuss the probable content of the messages themselves as they are received by the learner.

The final section of this chapter will be a summary of communication patterns and social learning in everyday life. This summary will suggest some of the unanswered questions that must be addressed by further research, and will speculate on the implications of our research for social policy.

THEORETICAL UNDERPINNINGS

Social Learning Theory [3]

1. Our Starting Point: Assumptions and Definitions. Social learning is used here to describe that process by which human beings in concert with their environment acquire ideas, affections, and behavior. The theory is to be distinguished from conceptualizations of learning and behavior that assume change is a product of some innate drive or internal physiological or psychological structure of the individual.

Social learning theory, at least in more recent formulations, differs from classical behaviorist approaches in that it does not exclude many of the concepts of cognitive learning theory. Recent social learning approaches can be labeled as S–O–R approaches. S and R stand for stimulus and response. O stands for the orientation of the individual to the stimulus; that is, the personal interpretation (conscious or unconscious) through which stimuli are filtered. This internal process may alter the message being communicated by the stimulus and may result in a modified or unexpected response. A further modification of classical behaviorist theory suggests that stimuli may be generated from within the person by thinking or feeling operations, thus producing a response without any external stimulation. This very important distinction made by social learning theory is definitely useful to our exploration of sexual learning and communication.

Any learning theory must address four primary issues:

1. What is learned
2. How learning occurs
3. Where learning occurs
4. The influence of the location on what is learned and on how learning occurs.

In social learning theory the "what" of learning concerns responses or response patterns and their links to particular stimuli. The "how" of learning is positive and negative reinforcement to particular responses to particular stimuli, modified by such factors as schedule, distribution, and cognitive orientation. The "where" of learning refers either to the external physical and social environment or to the individual's subjective reality. The important considerations about the location of learning are the way in which the environment is perceived, the persons or institutions who structure and organize the environment, the aspect of the environment that is mainly responsible for learning, and finally the constraints or limitations that the environment places on learning.

In the following section on how social learning occurs we discuss three main concepts of social learning theory: conditioning, reinforcement, and modeling. Important components of social learning theory are stimuli, perception, cognitive orientation, and goals and expectations. They will be discussed when they are relevant to these three concepts and to questions about the location of the learning environment. A stimulus is an event or a condition that elicits a response (nonresponse is considered a response). Perception is the process by which one absorbs stimuli through the five senses—plus the internal process that takes place when the cognitive orientation and sensory perception come together. Cognitive orientation is the framework, shaped in part by past experience, for interpreting stimuli and integrating current experience into behavior. Goals and expectations will be considered here as a subset of an individual's cognitive orientation and will be discussed in terms of their effect on responses.

2. How Social Learning Occurs. Although our concern is social learning in the complex situations of everyday life, many of the building blocks of social learning theory were established through experiments in special, controlled environments. While these circumstances differ from the more complex circumstances of real life, the simplicity and control of the experimental environment allowed the interacting and compounding aspects of the social learning process to be isolated and identified. From these experiments were developed the concepts of conditioning, reinforcement, and modeling. We will briefly review them here so that we may later explore how they come into play in a variety of situations.

Classical conditioning, typified by Pavlov's dogs, is based on a simple stimulus-response paradigm. The subject learns that a light or a noise is a signal that something pleasurable like food, or something painful like an electric shock, is on its way. The subject is able to prepare for what is coming. However, this is not an interactive system. What is coming is coming. The subject has no effect on the stimulus; that is, no effect on the environment. These conditions almost never exist in real-life interpersonal situations. Even in situations where there is a significant built-in power imbalance—such as parent-child, teacher-student, doctor-client—there is always a repertoire of behavior available to the subordinate (tantrums, angry looks, stubbornness) which does, in fact, affect the stimuli, the environment, and the outcome of the situation.

E.L. Thorndike and B.F. Skinner are probably the names most often associated with instrumental learning. Skinner put hungry pigeons in a box where they pecked around randomly. When they happened to peck on an illuminated window, a mechanism delivered a food pellet—a form of reinforcement. Eventually the pigeons learned to go straight for the window when they wanted food. In this model, the subject—pigeon or human—has a clear effect on the environment. By certain behavior the subject can elicit rewards—positive reinforcements from the environment.[4]

The process of reinforcement between humans in real life is often unintentional and may encourage undesirable behavior. For example, parents may unintentionally encourage children to whine by withholding their attention until the whining becomes too much to tolerate and the parents give in. Thus, although reinforcement may not always be easy to predict and control, it still has an effect upon the learner.

In our daily lives, of course, we are constantly bombarded with so many stimuli that we could not possibly respond to all of them. Some choosing, conscious or unconscious, is clearly necessary. Social learning theory says that prior learning guides our senses and determines what we notice and respond to. For example, in a 1961 study by Toch and Schulte[5] subjects were tachistoscopically shown violent pictures to one eye and neutral pictures to the other eye simultaneously. There were two groups of subjects—one made up of third-year police administration students and one made up of students just beginning their first year in the same program. Both groups were presented with identical stimuli, and yet the third-year students reported seeing much more violence than did the novices. Presented with a range of stimuli, the students responded to pictures of violence according to their prior training.

The strongest pillar of social learning theory is modeling, or learning by observation. Studies have shown that modeling can be effective in acquiring new response patterns. It can have an inhibiting or encouraging effect on present patterns, and in new situations it can elicit responses that are already in the person's repertoire.

Modeling works in various ways. Subjects have learned behavior, emotions, and affective expression. Models may be live or symbolic—though it should be noted that a live person, whether present or on film, is more likely to be modeled than a cartoon character or other graphic representation. Many studies have shown that a model is more effective if the subject feels a strong personal tie to him or her, or if the model is prestigious.[6]

Modeling does not simply refer to expressing behavior in the presence of the model. Subjects generalize what they have absorbed from the model to situations where the model is absent[7] and to situations that have not been specifically modeled. This has been demonstrated in issues concerning moral judgment[8] and in the development of patterns in which the individual learns to delay gratification.[9]

In most situations the principles of modeling and reinforcement operate in combination. One assumes that the extent to which an imitated performance becomes a lasting part of a person's repertoire will vary with the extent to which it is rewarded or punished in the learning situations and in later situations. We should also remember that the individual's self-image and the process of self-reinforcement (derived in part from previous social learning) also affect behavior. In fact, particularly in older children and adults, self-reinforcement may outweigh immediate external responses as a determinant of behavior.

There are many examples in the literature of how previous social learning can affect an individual's response to any situation. Studies have shown that girls, schooled in habits of dependence and non-autonomous self-concepts tend to be more susceptible than boys to the influence of modeling. However, in an aggression modeling experiment, boys—who presumably are allowed more leeway in this matter—tended to imitate behavior that girls did not.[10] This was true whether the model was rewarded, punished, or received essentially no response (although all of the children in the model-punished group, of course, did the least imitating). Later in this experiment, boys and girls in all three groups (model-rewarded, model-punished, no consequences to model) were offered powerful reinforcements for aggressive behavior. The author reports that "the introduction of these positive incentives for performance of aggressive behaviors practically wiped out the previous disparity between the sexes.[11] The author further suggests that the inhibitory effects of differing aggression reinforcement histories were clearly reflected in the finding that, during the reward phase of the experiment, boys were more easily disinhibited than girls. This is a significant study for those of us whose generation(s) are caught up in great social and sexual change. It is one of the few which may help us to predict what a person will do when confronted with a discrepancy between prior social learning and new conditions and messages.

The phenomenon of conflicting "lessons" is a key issue in any theory of human development and change. This may come about from conflicting messages received from different sources such as home and school, or "the experts" and individual experience. Or the conflict may arise from mixed messages from a single source. In 1966 Mischel and Liebert[12] conducted a study involving the latter sort of conflict on the adoption of patterns of self-reinforcement. Three groups of children participated in a game with a female model. In the first group the model rewarded herself only for very high scores but rewarded the children for relatively low scores. In the second group the model was lenient with herself but rewarded the children only for high scores. In the third group the model rewarded both herself and the children only for high scores. Later the children were left to play by themselves, and their patterns of self-reward were measured. The results were as predicted. The third group, having observed and received consistent strict criteria for reward proceeded to impose the same stringent measures on themselves. Both of the other groups, which had experienced examples of inconsistent behavior, went on to adopt more lenient standards for themselves. Differences, however, also separated these two latter groups. The group that had received stringent direct training and observed lenient self-reward showed much more conflict about rewarding themselves for mediocre performances than the group whose direct training had been lenient. Taking a small leap, it seems fair to say that parents of the do-as-I-say-not-as-I-do school have an increased chance of raising conflicted children.

In response to inner drive theories of human behavior we seek not to deny them utterly, but to say that they do not explain enough. Albert Bandura has said, "An organism that is impelled from within but is relatively insensitive to environmental stimuli or to the immediate consequences of its actions would not survive for long."[13] In fact, the history of humankind provides great evidence that we are organisms that have had to adapt to a wide variety of conditions—those of our own making and those created by other forces. As a species, we are certainly among the survivors.

3. Where Social Learning Occurs. The important questions about the location of social learning are: (1) in what way is the environment perceived, or, how is functional meaning attributed to it? (2) who organizes and structures the environment? (3) what factors in the environment are most responsible for making learning take place?

The environment can be perceived as the source and place of all learning. Stimuli and reinforcers all emanate from the environment, and behavior is defined by its occurrence in the environment. Antithetical perspective describes the environment simply as the stage on which behavior is enacted. Behavior in the second description is the result of other, more intrapersonal, forces. Social learning theory takes the former of these two positions. For example, it assumes that when an individual learns dating behavior, the stimulus arises from a source such as noticing one's peers dating; the reinforcement may come from satisfaction with the experience or social approval; dating behavior means what one does in dating.

The identification, organization, and definition of stimuli and reinforcers in the environment is very important to learning. One viewpoint is that the learner is the one who performs the acts; another viewpoint is that this performance is done by forces external to the individual. Social learning adopts the former viewpoint. For example, in learning about dating, the learner determines what dating is, who she or he wants to model, and what is positive and negative reinforcement for dating.

The question of what is responsible for learning also raises two opposing views. One is that the stimuli in the environment—along with the inner states and needs of the individual—mainly determine what learning occurs. The other view is that the reinforcers in the environment—along with the individual's cognitive orientation—mainly determines what occurs. Social learning theory adopts the latter viewpoint. For example, in terms of dating behavior, what is learned is mainly what leads to "success" in dating.

Communication Styles

A person receives, makes sense of, and responds to messages that come either from other people or from the structural features of his or her environment. These messages are transmitted through many channels. In human communication they may be intentional or unintentional, verbal or nonverbal.

1. The Spoken Language. Spoken language plays a huge, and usually ignored, role in social learning. According to Benjamin Whorf:

> We dissect nature along lines laid down by our native languages. The categories and types that we isolate from the world of phenomena we do not find there because they stare every observer in the face; on the contrary, the world is presented in a kaleidoscopic flux of impressions which has to be organized by our minds—and this means largely by the linguistic systems in our minds. We cut nature up, organize it into concepts, and ascribe significances as we do, largely because we are parties to an agreement to organize it in this way—an agreement that holds throughout our speech community and is codified in the patterns of our language.[14]

In the very process of language acquisition a child is partially initiated into the world view of his or her community. As adults we can neither conceive of, nor communicate about, concepts for which we have no name. Basic experience, such as the sense of time and space, is largely determined by language.

It is still useful to recall George Orwell's examination of "newspeak,"[15] the manipulation of language for political purposes. Certainly Vietnam war reports would have been more upsetting if we had been told that our army was murdering farmers and their families rather than introducing limited use of antipersonnel weapons; burning the skin off people rather than napalming the area; destroying the crops, land, and water supply rather than defoliating. These words were used to distance us from those events, and they did so even though we believed that we saw through the process. Similarly, a child may have his or her sense of reality invalidated by hearing from a parent: "Mommy and Daddy are not fighting. We are having a discussion." When the child has seen angry faces and heard loud voices, this assertion becomes confusing, to say the least.

John B. Carroll and Joseph B. Casagrande explored the question of whether linguistic differences would be carried over into nonlinguistic behavior.[16] We will deal here with only one example from their study of Hopi and Anglo adults. There are many differences between the Hopi language and English. One of these is that the Hopi "uses the same words for spilling and pouring, but must use a different verb depending on whether the material being spilled or poured is liquid or non-liquid."[17] Anglo and Hopi subjects were shown the three drawings below (Figure 2–2) and told to say which two pictures went together. The dominant Hopi response was to pair A and B, which involve solid materials; the dominant Anglo response was to pair B and C. Furthermore, the authors report that a number of Anglos asked whether the man in picture A was pouring or spilling the peaches. Not one Hopi asked that question.

Carroll and Casagrande conclude from this distinction and from the rest of their study that linguistic differences do in fact have nonlinquistic behavioral complements. Research leaves us uncertain, however, about exactly how far-reaching the implications of these findings are for learning. If English has more words that distinguish between purposeful and accidental happenings (more words by which to label something a mistake), are Anglos then more preoccupied with accidents and mistakes? If we drop something, do we feel more shame or embarrassment than the average Hopi?

Figure 2–1. Examples of Spilling and Pouring from Carroll and Casagrande Experiment.

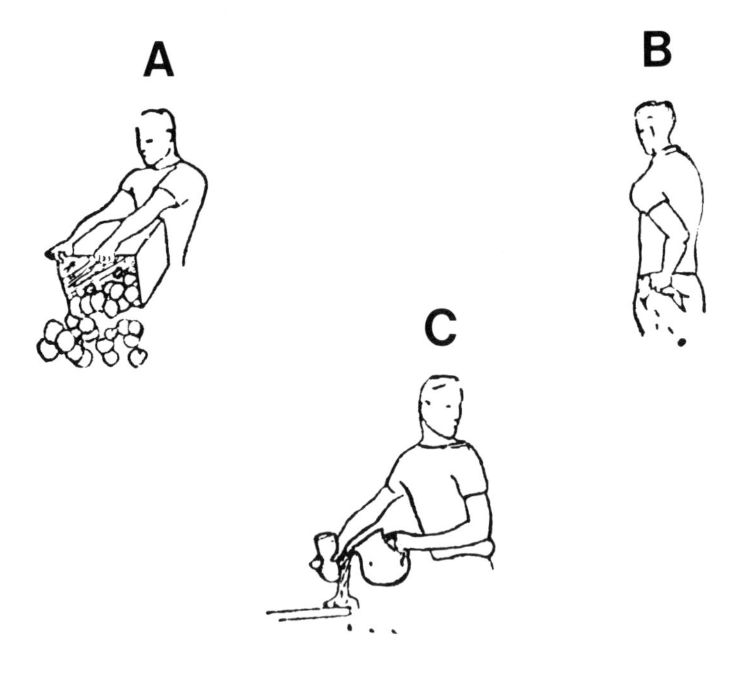

Whatever the answer to these questions, it is certainly true that language plays a large part in forming our ideas and values. It is no wonder that in its early stages the women's movement quickly pointed out some of the sexism inherent in our use of the English language. The use of the terms man or mankind to refer to all people has been shown to be part of a linguistic system that posits the male as primary subject and the female as "other." Furthermore, distinguishing a "waiter" from a "waitress" contributes to our notion that work is sex-defined even where there is no real distinction to be made.

Robin Lakoff, in *Language in Society*, discusses both the concept of "woman's language" in English and the deprecation of women implicit in general usage.[18] Examples of woman's language are given in lexicon, syntax, intonation, and other suprasegmental patterns. Lakoff illustrates the weakness introduced into woman's English by its use of tag-questions ("The Vietnam war is terrible, isn't it?") and tag-orders ("Won't you please close the door?"). Other examples can be added to this list: hesitating, apologizing, and disparaging one's own statement ("I don't know anything about it, but . . . ").[19]

On the one hand, these linguistic habits are a manifestation of women's social learning. Women have learned that what they have to say is not important. Women have learned not to be too forceful and not to trust themselves. On the other hand, women's English is also a part of the social learning of men. Through this style of speech, women convey, legitimate, and reinforce an attitude of disrespect toward themselves and other women. Maija Blauberg has found these different male and female language styles to be reflected, perhaps exaggerated, in children's literature. Children's books typically have women asking more of the "what do you think we should do?" sort of questions. They also show women frequently saying such supportive things as "gee" and "that's great."[20]

Words are only one aspect of language. It is not only what we say but how we say it that counts. Girls and boys not only use different phraseology, but have different intonations as well. J. Sachs found that subjects could easily distinguish between the voices of boys and girls despite the fact that the children were young enough so that there was no physical reason for a vocal difference to exist.[21]

A number of studies have shown that mothers tend to talk to girl infants more than to boys. Parents also exaggerate their responses as a way to show their children how to respond to various situations—that is, how to feel. Cleason tells us that parents tend to say things like "hey, wow, that's almost full to the top," while simultaneously modeling, one would presume, correct facial affect. This is not how one adult speaks to another adult. In general, Cleason found that the language that parents use in speaking to their children is often a purposeful language of socialization.

Adults transmit and receive social messages through the more subtle aspects of language. For instance, forms of address are strong carriers of status messages. It is commonly accepted that a dentist may refer to his dental hygienist as Mary while she is expected to call him Dr. Jones. This example is, of course, compounded by sex differences, but the rule also holds for people of the same sex. Dan Slobin, Stephen Miller, and Lyman Porter[22] found in their study of address and social relations in a business organization that people tended to be "more self-disclosing to their immediate superiors than to their immediate subordinates." Similarly, Erving Goffman observed that "the boss may thoughtfully ask the elevator man how his children are, but this entrance into another's life may be blocked to the elevator man, who can appreciate the concern but not return it."[23]

Many of these patterns are also replicated within the family. Parents tend to be addressed by role title (Mom or Dad), while children are addressed by first name. In the family and in society at large, language is both a reflection and a source of learning about power relations and all other aspects of social life.

2. Nonverbal Communication. People "speak" to each other not only with words and sounds, but with looks, gestures and actions. In recent years more attention has been paid to various aspects of nonverbal communication. Nonetheless, Nancy Henley points out the interesting paradox that while we are all quite familiar with nonverbal cues, this language has never been formally legitimated.

> Our culture emphasizes verbal over non-verbal communication. English is taught in our schools through all grades, with the aim of better understanding (diagramming sentences, learning Latin roots) and better expression (writing compositions). Non-verbal communication isn't taught: we never learn to analyze what certain postures, gestures, and looks mean, or how to express ourselves better non-verbally. (Of course, non-verbal communication is learned informally, just as language is learned before we enter school to study it.) This doesn't mean that everybody doesn't *know* that looks and postures mean something, perhaps everything, especially in emotion-charged interaction. But mentioning looks and postures is illegitimate in reporting communication; legal transcripts and newspaper accounts don't record them. And they are seldom allowed in personal argument ("What look? What tone of voice? Did I say OK, or didn't I?")[24]

Michael Argyle and his colleagues have estimated that a nonverbal message carries 4.3 times the weight of a verbal message.[25] Whatever the exact ratio may be, nonverbal communication is real and must be included in any thorough examination of the processes of communication and social learning.

There are many forms of nonverbal communication, and lack of expression may say as much as full expression. We offer or withhold support by touching or not touching, by coming close or maintaining our distance, by being available or unavailable. In terms of social learning, nonverbal communication is one way that we learn. If, for example, a child grows up in a home where masturbation is not discussed and where she or he never sees anyone masturbate, the child receives this as a message. If a child asks his or her parents where babies come from and the parents "freeze up" or sweat or fidget, the reaction is a message. Even if the parents go ahead and give a perfectly adequate and articulate verbal answer to the question, according to Argyle's calculations the main message will be that mentioning sex makes people (parents) nervous.

Some body responses, like blushing, are seemingly universal; others are more clearly culture-specific, and misreading them can make for trouble. For instance, recently one of the authors was in an elementary school and happened to observe an Anglo teacher who had brought a Puerto Rican boy into the hall to bawl the child out. The boy was listening to the teacher with his head lung low, eyes to the ground, in what the child had learned was a posture of respect. The teacher, coming from another culture, apparently decided that the child was not paying full attention and was therefore being disrespectful. Grabbing him by the chin and pointing his face toward her own, the teacher shouted, "Look me in the eye when I'm talking to you." Such an incident is a clear case of a misunderstanding based on different social learning about body posture and respect.

Some body postures are manifestations of or responses to past social learning. An individual may continue to exhibit these postures despite the fact that there is no immediate external stimulus that calls for such a response. The stimulus at this point has become internal—it is part of that individual's cognitive orientation. These stimuli may be difficult to document or to describe, but therapists such as Ida Rolfe and Wilhelm Reich have observed and written about them for years. Freud, of course, postulated that experience and knowledge that were too difficult for an individual to handle were repressed and thereafter resided in the unconscious. The experiences and the feelings that accompany them do not disappear; they remain to exert indirect effects. Reich, Rolfe, and others have gone on to say that these feelings or traumas become essentially lodged in the body, in our musculature, in our breathing, and so forth. In the case of a woman who in her youth was made to feel embarrassed about her developing breasts and consequently developed a chronically caved-in chest and rounded shoulders, we would say that this is a result of her social learning about her sexual self-image.

Most doctors, who are often considered body specialists, do not read body language or attend to these cues. This, as does the fact Henley observed—that official communications rarely include such information—teaches us to ignore these aspects of our own experience by invalidating them through neglect. We are taught to separate mind, body, and feelings, and to trust the mind above all else.

The Allocation of Resources as a Form of Communication

In addition to verbal and nonverbal modes, the allocation of such resources as time, space, energy, and money is the source of much social learning. Since much of this communication may be unplanned, it is best understood when examined from the angle of the learner or the person who is receiving and deciphering messages. These messages may come from other individuals, from social institutions such as the mass media, and from such semi-intangibles as the organization of life in a given culture. Some examples should make this point more clear.

1. Time. The Basic thesis is that time is essentially a medium of exchange. We learn what activities are important or unimportant partly by how much time is devoted them. As one aspect of modeling, the use of time may compound or conflict with any verbal messages. "Time talks," writes Edward T. Hall in *The Silent Language*. "Because it is manipulated less consciously, it is subject to less distortion than the spoken language. It can shout the truth where words lie." To illustrate this point, Hall describes his own encounter with several government officials and the dramatic discrepancies that he observed between their verbal commitment to an issue and their actual apportionment of time.[26]

Comparable uses of time as a reflection of the relative value of a person or activity can be seen in such learning environments as the family and the mass media. For instance, people demonstrate a particular relationship to their bodies by how much time they devote to exercising, massaging, bathing, and feeding them. Although time is only one aspect of this relationship, it is an important one. Moreover, it is something that the child will learn from the parent. Messages about the value or importance of an activity are also communicated from the social and structural environment to the individual or family. Examples here would include the time requirements of mandatory schooling or the standard eight-hour work day.

Conceptions of time vary from culture to culture. In our culture, for instance, "time is money." Time is to be saved, or spent wisely. In production the emphasis is on efficiency, and this attitude filters down to other aspects of life. Consequently we are not a nation of leisurely diners. We rush to fast food outlets and fast car washes; we value the ability to think on our feet or provide a quick comeback; we believe in love at first sight and count orgasms.

2. Space. Space is also a vehicle for and an aspect of social learning. Frequently, the dimensions of time and space intersect. Svend Reimer found that, within the population he studied, the smaller the family apartment was the greater the tendency of adolescents to spend evenings outside the home.[27]

The ways in which we design and use space reflect our values. It is true, too, that the characteristics of the spaces in which we spend our time influence us and our value system. As always in complex social situations, it is difficult to isolate the effect of any single variable. An example from one of the authors' own experience may help, however, to illustrate the point.

Recently an analysis of the use of space was conducted in a federally funded housing project in Worcester, Massachusetts. Both the design (that is, the formal use of space) and the result (that is, how the environment was actually experienced and used by the residents) were studied. In an interview, a group of fourteen-year-old white boys were asked where they went to be alone—alone with a friend and alone with a girl friend to make out or make love. In response to the latter question, they said that—depending on the weather—sexual activity went on either in the stairwells of the high-rise buildings, or on the baseball field (the one recreational space allotted to adolescents in this vast project). When asked if the likelihood of invasion was a problem, the boys laughed, jabbed one another's ribs, and said, "Well, you've got to do it somewhere. We all know everyone does it, so who cares if they see me? You should have seen his brother and Mary Sue on the ballfield last Tuesday. We was all watching them do it." Clearly, the attitudes, values, and behavior patterns that go along with making love in a dark, cold stairwell that is vulnerable to intrusion, are different from those that go along with lovemaking in a room that is warm, light or dark, and controlled for privacy.

Commanding space is one aspect of power. This corresponds to a conception of space as the individual's or family's territory—their sphere of action, existence, and thought. Most wars have been fought over the control of space. As Dr. Chester Pierce[28] has observed, the assumption or abdication of space occurs every day on sidewalks and at doorways where one person must give way to another. Consistently, blacks defer to whites, and women defer to men despite the rules of chivalry. Similarly, Pierce has pointed out that the class nature of various sports is revealed through the occupation of space. Lower class sports often are played in the streets. Even when played on official courts or fields, sports such as basketball require much less space per person than tennis, polo, and other games of the elite.

Individual and social values are reflected in the use and control of space. Something about a person, for instance, is reflected simply by the fact that she or he feels the need for a study, or does not need a dining room in the home. On a political level, basic values and power relationships are made manifest when new intercity highways are routed through a poor neighborhood. The disturbance to community patterns and family roots makes a clear statement not only about who controls space, but also about the inevitable effect of this control on the lives of those who don't have it.

3. Energy. A third resource whose allocation can be understood as a channel of communication is energy. For a number of reasons energy is more difficult to discuss than either time or space. Yet, the discussion is warranted by the impact our perceptions of each other's energy levels and expenditures have on our interpersonal relations and thereby our social learning.

The first problem is that a useful discussion of energy as a means of exchange between humans demands some agreement on exactly what we mean by the term "energy." Formal definitions and colloquial usages abound. To some, energy is the force it takes to move an object a given distance at a given speed; to others it is what is produced when calories are burned; to others it is what young kids have too much of and older folks pine for (with the clear indication that this involves something more than movement and caloric consumption). The American Heritage dictionary defines energy as both "vigor or power in action" and "vitality and intensity of expression."[29] These descriptions come closer to the concept of energy that is helpful in discussions of human communication.

When speaking of human energy as a resource, we are speaking of the attention that one has to give and chooses to give to a person, an object, or an activity. One critical problem is how to measure this attention or energy. Whereas time and space are more tangible resources with agreed upon units of measurement, there are no such agreements about human energy.

One aspect of energy that can be measured, of course, is the time period during which it is expended. How much time during an average week does Mr. Jones spend paying attention to his son, Billy? How much time does Mr. Jones spend playing his cello? Watching TV? Since time usage as a medium of communication has already been discussed, we can temporarily disregard that aspect of energy. We are left with the problem of identifying and measuring the other aspects of energy.

An important aspect of energy that we measure frequently is, for lack of a better word, intensity. This is, essentially, the difference between divided and undivided attention. Mr. Jones may spend a lot of time paying attention to Billy, but if while Billy is talking to him Mr. Jones has half his mind on an unresolved problem with a coworker, then Billy is not receiving a lot of energy, and he will know it. He may make sense of it in many ways. He may think his father finds him boring or donesn't care about him; he may think his father is getting old and fuzzy-minded. We don't know what Billy will think or feel, but we do know that he will notice.

No formal measurement units exist for the intensity of human energy, but our subjective measurements are often stated in the language of proximity or space. We may say that a conversation partner was "right there" or "far away" depending on our sense of the portion of attention, or energy, that we felt was devoted to the conversation.

Our social learning is derived partly from the kind of energy we receive from others; partly from the kind of energy we see significant others devoting to various activities or people; and partly from the kind of energy we devote to the people and activities in our lives.

It is also true, of course, that our social learning in large part determines how we expend our energy, where we give our attention. For instance, many black people are much more knowledgeable about white people than vice versa. Blacks often know and can imitate the mannerisms, habits, attitudes, and life styles of whites in a way that most whites do not reciprocate in relation to blacks. The same can be said of women in relation to men. Blacks and women have paid a lot of attention to the patterns of whites and men because learning to understand and predict people who have power over them may give them a measure of control over their own lives. They have felt it an important way to spend energy.

The ways in which we use energy, then constitutes messages. They are borne or our prior social learning, and they contribute to the social learning of those around us.

4. Money. The allocation of money transmits a very straightforward message. Children absorb parental values (and citizens absorb social values) partly by experiencing where and how money is spent. In terms of social learning, it should be remembered that these expenditures are not simply a matter of single messages—statements made only at the moment of decision or payment. These messages are continual. For instance, when a family decides to buy a television set rather than a few bicycles, they are making a statement about their priorities. That statement remains with them every day in the presence of the television and the absence of the bicycles, and comes to determine how they will spend their time.

In the purchase of commodities for personal use we are dealing, in effect, with our own direct and indirect influence upon our own lives and learning. In a society where we have less and less influence over most aspects of our lives, the potential power of our monetary decisions becomes increasingly important to us. This leaves us quite vulnerable to the distortion of advertisements. It used to be a popular cliché that "money can't buy love." According to most advertisements, this is no longer true. Love (or sex) seems to be a fringe benefit that comes with most cars, cigaretts, and deodorants. You can buy love, and you can buy peace of mind, excitement, and respect—you can buy being the neatest kid on the block, too. Our own money is one of the few resources with which we consciously (and unconsciously) try to express ourselves and shape our lives. Most other realms are beyond our control.

The allocation of money both reflects and perpetuates the social order. The wage system in this society is clearly stratified. Women and minorities are paid less than white men. Labor receives less than management. It is not a system based on individual need, capabilities, or even the real value of the work. Clearly, the work of a nurse's aide is as indispensable to a hospital as the work of the chief administrator. In a society based on money, this payment structure is an essential determinant of the quality of one's life. This system transmits strong messages about what one is entitled to and therefore what one is worth in the financial and human senses. Such statements of "objective worth affect one's inner sense of worth.

We are not compartmentalized beings. Lessons learned in one realm spread to other realms and become part of the total mix that makes up our sense of ourselves, our world, and our place in it. As time passes, of course, we lose track of the source of such messages and are left with only the messages themselves. Having forgotten where these self-images originated, we read them as truths about ourselves. If we care about what people come to think of themselves, and if we care to alter what for many are negative self-images, then we must trace our way back to these sources. That is one underlying purpose of this work.

LEARNING ENVIRONMENTS

Introduction

In the sections that follow, the discussions are informed by the ecological approach to human development articulated by Urie Bronfenbrenner.[30] According to this approach, human development takes place within three concentric layers: the immediate setting, the institutional context, and the ideological system. The immediate setting is the world closest to the child—that is, home, school, street, and playground. Surrounding these are institutions such as the health, education, and welfare services that have direct responsibility for the child. Included here, too, although sometimes seeming more distant, are economic and social planning agencies, industry and business, and the mass media. Finally, wrapped around both inner layers is the ideology that informs and directs the institutions and the immediate setting.

Bronfenbrenner's analysis suggests dimensions to the depth of each of the environments described in the sections below. In order to see this depth as well as the interrelatedness of the environments, it may be helpful here to take a brief look at the historical process that lies behind two of them: family life and work life.

As the centerpiece of all the learning environments, the family has always had and still has primary responsibility for the care and socialization of children. Although many of the other environments (media, peers, the social welfare system) share that task, this sharing has not always been a mutual decision, and is often experienced by parents as an intrusion. Thus, the family often finds itself in a tense or antagonistic relationship with schools, playmates, the mass media, and even with the experts to whom they turn for help.

The family's second major function is to be a haven from the outside world—a place where, more than anywhere else, we expect to receive love, appreciation, security, validation, and fulfillment. This monumental task comes in reaction to a historical process.

Prior to the eighteenth century, family life and work both occurred in one place. In the form of domestic and family industry, the home was the site of production. Although men and women tended to do different work, all labor was directly related to the family and absolutely vital to the family's survival. Children, too, worked, either in their own homes or in some sort of apprenticeship.

The family, then, was considered and experienced as both a productive and an emotional unit. According to Eli Zaretsky, "[the family's] members understood their domestic life and personal relations to be rooted in their mutual labor."[31] The sharp distinction between home and work that we now assume did not exist.

Physical survival, emotional care and growth, on-the-job training or job preparation, and whatever schooling existed were all a part of the social learning that took place in the family environment. Additionally, skills that were largely women's, such as healing, took place within the family or immediate community and were learned through observation.

The eighteenth century saw the transition from family and domestic industry to capitalist industry—which in turn paved the way for the development of today's educational and social institutions. When work and home separated, men's and women's work became even more differentiated because labor now existed in different locations and was seemingly related to different spheres of life. Working in the outside world, men became responsible for production. Women became responsible for the realm of "personal life," a new entity that was centered in the home. Zaretsky points out that, whereas self-cultivation, self-examination, and physical and mental development had previously been restricted to the leisure classes and to artists,

> Under capitalism an ethic of personal fulfillment has become the property of the masses of people, though it has very different meanings for men and for women, and for different strata of the proletariat. Much of this search for personal meaning takes place within the family and is one reason for the persistence of the family in spite of the decline of many of its earlier functions.[32]

Thus, two environments, the family and the workplace, were created where one had existed. Our expectations of each and our experiences with each are very different. Consequently, each provides for a distinct kind of social learning. Despite their divergent purposes, our examinations of these and of the other four learning environments will involve the same principles. We will look at what is learned through verbal and nonverbal communication, and through the allocation of space, time, money, and energy. We will explore physical space, and the things, human relationships, and activities found in this space.

Learning in the Family

The family in this society has two major functions. Not only is it the primary source of love and validation, but it also has the chief responsibility for the socialization of children. Clearly the two sides to this role are often antagonistic. The family is supposed to be different in important ways from the very environments for which it must prepare its children. Also, in a sense, parents must prepare their children to live in a world that the parents have never known.

Added to the difficulty of balancing these two tasks is the fact that in its role as a socializing force, the family occasionally finds its own authority usurped by outside agencies. Thus, many parents may be ambivalent about any help offered to them. They want the services and attention for their children that they cannot offer themselves. However, the tension arises when the source of support (school, YWCA, etc.) is providing not only neutral activities, but also a set of values or a world view that may be in conflict with the values of the home.

During the first few years of life a child has relatively little contact with the world outside the family. Then, through television, school, church, and friends, this contact and its influence steadily increase. Nonetheless, as the main source of a child's first social learning, the family has a profound influence on the development of the lens through which the child sees and understands, lets in or screens out new experiences and stimuli. Although social learning is a lifelong process and people can grow and change at any age, the formative social learning that goes on within the family is very tenacious. We expect to be loved or fought with in the same ways and for the same reasons that we were in the family. Later experience that alters or supplants this early learning normally has to be very potent or often repeated.

As in other environments, social learning occurs in the family through modeling and reinforcement. The most common form of familial modeling takes place when parents and older siblings are modeled by younger children. Reinforcement among family members is expressed not only through words and gestures, but through the common allocation of household time, space, energy, and money.

> In whichever way we attempt to realize them, we all seek certain goals: (1) affect, that is, intimacy and nurturance—that sense of loving and being loved by someone in our world; (2) power, the freedom to decide what we want, and the ability to get it—whether it be money, goods, or skills; (3) meaning, or some kind of philosophical framework that provides us with explanations of reality and helps define our identity. . . . This then is our basic thesis. Members of family gain access to targets of affect, power, and meaning through the way in which they and their families regulate the media of space, time and energy.[33]

The use of household space transmits messages about individual characteristics, about internal family dynamics, and about the family as a unit in relation to others. The first experiences with privacy and invasion come in the home, and probably have implications that affect the sense of self, relationship to others, and even concepts of political freedom. In this regard, studies have shown that crowded living conditions can make children more aggressive in school settings, contribute to family problems, and encourage a poor self-image.[34]

Family members let visitors know in which physical areas they are free to roam, just as they let outsiders know in which interpersonal and conventional areas they are not welcome, and which questions may be asked and which are "out of bounds."[35] Limits may be conveyed by not inviting a guest into a personal space or by comments like "have a seat in the living room while I go fix us some coffee."

There is no one-to-one correlation between use of space and the message that a person would internalize from it. One can only speculate about such learning and suggest that further research could be highly informative. If a child grows up in a household where each child has his or her own bedroom but the parents share a bedroom and have no personal work or study space, what does that mean? Does the child assume that the parents have no need for separateness? Does it contribute to the parents being seen as a unit rather than as individuals? If one family has regular seats at the dinner table and another family sits at random, what does it mean? Does it correspond to a general trend toward or away from routinization, or is it an isolated feature in their lives? Does it provide a good sense of security, ownership, and power, or is it needlessly territorial? Do members of a household freely join one another in bedrooms and bathrooms or not? Do they talk about their bodies or not? Do they share their naked selves or only their clothed selves?

Among the numerous daily learning incidents centered around the family can be found two kinds of general patterns. On the one hand, there are patterns in the tone or style of household that affect all family members in roughly the same manner, making each family a kind of parent-child communication environment. Within this environment the family's real influence is on the individual's basic orientation to life rather than on the specific content of family messages about various topics.[36]

On the other hand, there are also patterns of family interaction in which each member has his or her role to play, thus encouraging individuals to develop differently.[37] Each family develops certain strategies for decisionmaking, conflict resolution, and other family affairs in which family members, like members of an office work group or any other social unit, play standard roles. There are challengers and appeasers, intellectualizers and emotionalizers, and so on. Members may sometimes act out of character, but usually they do not. Such strategies are often carried outside the home, too, and applied to nonfamily situations.

Important social learning also goes on between siblings through talking, playing, fighting, helping each other, and watching one another in interaction with parents and friends. Studies have shown, for instance, that young siblings do their early sexual exploration together by exploring their own and each other's bodies and that children who have opposite sex siblings tend to display more cross-sex behavior than those who do not.

Some of the least predictable social learning probably arises from the conflicting messages that abound in most homes (as they do in most other social settings). The conflict may come when verbal messages don't correspond to nonverbal messages. For instance, a child may be told time and again that she or he is deeply loved and very important to the parents, but if dinner lasts half an hour and only two minutes of it are spent discussing the child's day or concerns, then the verbal message may be hard to believe. Or conflict may arise when a particular member has rewarding experiences outside the home that clash with family values. Eradicating all conflict from our lives is clearly an unrealistic goal. In fact, it may not even be a worthy goal since too much homogeneity can be boring or even deadening. Recognizing that conflict is inevitable, perhaps we can teach ourselves and our children how better to manage it.

Family members are not the only elements of home life. We will discuss the impact of mass madia in a separate section, but it is worth reminding ourselves here that the presence or absence of books, magazines, radio, and television, with their specific content messages, are a significant aspect of home life. Moreover, whether mass media activities are solitary or social is an important family pattern. Similarly, toys, hobbies, and home projects are an area of social learning within the home. Hobbies and play can give not only varied experiences of sharing and solitude but also difference experiences of one's body and its potential for pleasure, creativity, industry, and pain. How a child plays and what a child plays with make a difference. Finally, the family will affect the extent to which one trusts and is therefore open to learning from the mass media, experts, the man on the street, and all other potential sources of input.

In summary, we can say that home life affects all levels of social learning—the learning of attitudes, values, and behavior. It may also affect every sort of relationship that a person experiences—to self, other people, the world at large, and even to things like books, trees, art, and work. The learning comes from interaction with other household members, through conversation, modeling (observation), reinforcement (the giving or withholding of love and respect), and through the allocation of time, space, energy, and money. Families differ not only in how they use the resources they have, but also in the extent of those resources. Some families have large houses and yards and much separation from neighbors. Others have small apartments, surrounded by asphalt or other buildings and can hear and be heard by neighbors. Some parents can provide adequately for the family's material needs and still have a good deal of time that they may or may not choose to spend with the family. Other parents and other children don't have that choice. These aspects of family life, while a result of the larger social structure, exert their influence on home life and the social learning that takes place there.

Learning in the Workplace

There are two basic, widespread, and quite opposite notions of the meaning of work and its role in a person's life. On the one hand, work is viewed as a potential arena for self-actualization. In contrast to this notion many Americans experience work as something distasteful that has to be done; it is the opposite of play, which is both voluntary and pleasurable. Nonetheless, whether enriching or frustrating, work for most adults (housework or paid employment outside the home) is the organizing feature of their lives and a powerful source of social learning (see Chapter 3). In all but the very highest and lowest strata of society, work is the only route to money, and money has everything to do with our level of material comfort and with the options available to us in our nonwork or leisure time. Moreover, our experience of work also has much to do with how we spend that leisure time, at least insofar as working may exhaust or energize us, make us feel good or bad about ourselves, excite or numb us, and encourage us to think or not to think.

As they grow, children are socialized for the world of work by seeing the adults around them working and by school and home preparation, both in specific skills and in a general knowledge of such things as competition and rules. In a sense, though, there is no real preparation for entry into the work force and this belief is embedded in our language. We speak of "the real world," meaning the world of work, the world outside the home. Because work is believed to be different from the rest of life, social learning at the workplace may supplant earlier learning. One enters a work situation expecting to find new rules and new lessons and is therefore open to new learning.

This is not to imply that either for the individual or society no pain or struggle is involved in this transition to the work world. E.P. Thompson's classic work *The Making of the English Working Class* describes a historical transition in which an entire generation of people was moved from agriculture and domestic industry into the early factories of England. People whose bodies and minds were accustomed to cycles that correlated with the seasons and with the rising and setting of the sun were reshaped into people who made their bodies and minds obey the clock.[38] For an individual, this sort of painful adjustment, along with the positive changes, all become part of the social learning at the workplace.

Since 1900, the data on employment in this country indicate a trend first from agriculture to blue-collar or factory work, and more recently, a decline in blue-collar employment and an increase in white-collar jobs. Setting aside professional work (which accounts for a very small percentage of the employed), we will chiefly examine factory work and office work.

Most workers have little or nothing to say about the allocation of time, space, energy, and money, as described in Part I. This is true despite the fact that it is mostly their time and energy being used. The overriding value behind these decisions at the workplace is cost-effective production. In most cases, human needs and desires for fulfillment, purpose, and interaction either do not figure into these decisions at all, or are considered as problems to be overcome.[39]

Laziness and sabotage exist, but they are less a comment on human nature than a response to the working conditions. The following examples come from the Vega plant in Lordstown, Ohio and the Fair Plan Insurance Company,* but they could be any factory and any office.

> While her son and his friends talked shop—DLOs, strike, rock bands—I talked with Mrs. Giusio in the kitchen. Someone in the supermarket where she worked had said, "Those young kids are just lazy." "One thing, Tony is not lazy. He'll take your car apart and put it together any day. . . . And I'm not lazy either. I love to cook. But supposing they gave me a job just cracking eggs with bowls moving past me on a line. Pretty soon I'd get to a point where I'd wish the next egg was rotten just to spoil their whole cake."[40]

> "The other day when I was proofreading endorsements I noticed some guy had insured his store for $165,000 against vandalism and $5,000 against fire. Now that's bound to be a mistake. They probably just got it backwards. I was just about to show Gloria [the supervisor] when I figured,

*Perhaps an assumed name.

> 'Wait a minute! I'm not supposed to read these forms. I'm just supposed to check one column against the other. And they do check. So it couldn't be counted as my error.' . . . If they're gonna give me a robot's job to do, I'm gonna do it like a robot."[41]

Apart from sabotage and passive resistance, others respond to the tedium with fantasies. In the Vega plant one worker told the interviewer:

> "I have fantasies. You know what I keep imagining? I see a car coming down. It's red. So I know it's gonna have a black seat, black dash, black interiors. But I keep thinking what if somebody up there sends down the wrong color interiors—like orange, and me putting in yellow cushions, bright yellow."[42]

If, in response to your work environment, you let yourself be half-dead all day, you may find it difficult to re-engage, to become fully alert and alive at the end of the day. The numbness becomes habitual. On the other hand, if you remain fully alive and sensitive to your surroundings, then the boredom and the physical pain of sitting or standing in one place all day will become unbearable. Social learning in many jobs, therefore, includes learning to turn off one's mind and to deny the body's pleading.

Some of the implications for sexual learning seem clear, although they would be difficult to research. It is hard to imagine that a person could spend his or her day divorced from his or her body and emotions, divorced from the sensual world, and storing up numbness and resentment, and become a tender, sensitive, joyful, finely attuned lover at night.

If you are told at work that you can be replaced by a machine, or by absolutely any other person, it is hard to feel unique, proud, or worthwhile. One may absorb that estimate of one's worth and feel awful or one might fight desperately against it to establish one's worth and uniqueness by putting other people down, by investing a major portion of one's resources in a customized car, or by a hundred other indirect means.

Before we leave this working class world of work, let us look at the human relationships—with superiors and with peers—that the job provides. In the company of supervisors, the worker and his work are in a position to be judged. In production jobs, especially large scale ones, evaluation by supervisors must be strictly in terms of speed and quantity, rather than by any human qualities. Any relationships that manage to form in spite of the hierarchy will necessarily be fraught with the ambiguities of this connection.

If you spend all day in service to the values of speed and quantity those values seep into your behavior patterns even when you may not mean to take them as your own. Without alternative experience, it may be difficult in other situations to care less about the product and more about the process. As a culture, therefore, our lovemaking is said to be characterized by speed, orgasm-counting, and performance fears. Where in our daily lives would we learn to enjoy a slow, sensual, nongoal-oriented process? The workplace values described above are often reinforced in school, sports, and the mass media.

Seemingly, more potential exists for human interaction between workers who are peers, and the relationships with one's peers at work has the potential to make the workplace more enjoyable. However, in many factories, the environment—if not the worker's superiors—prevents even the smallest exchanges between workers. A woman who worked on ping-pong paddles at Paragon Table Top Sports Corporation said,

> "You couldn't talk either. You wouldn't want to anyway because it was too noisy and the way we were all spaced apart, you'd have to lean over and shout. But if you even tried to talk Alma would come running over with a lot more paddles or she'd yell 'Why aren't you finished yet?' So you were alone with your head all day."[43]

In most cases, both the "demands" of efficient production (such as speed and noise) and the rules mitigate against personal contact between workers during work time. Only during the timed and regulated lunch and coffee breaks are workers able to make contact, and only then do any of the typical aspects of social learning among peers take place. The central theme of Christopher Argyris's book *Personality and Organization: The Conflict between the System and the Individual* is that the demands of factory work prevent maturation of the self and in fact require that an individual remain, or at least act, childlike.[44]

Argyris's observations may provide a partial explanation for Jane Loevinger's findings on the ego development of American adults.[45] Loevinger has described six basic stages of ego development, progressing from presocial infant to integrated adult. The data indicate that adults in this country cluster heavily at the conformist level (stages 1 to 3) and somewhat at the conscientious level (stages 1 to 4) with a sharp drop after that, thus exhibiting values and behavior appropriate to the work situation we have explored.

Obviously not all work situations are as dismal and limited as those described thus far. For instance, doctors, lawyers, news reporters, auto mechanics and musicians all have some control over their time and thus may be spared an education in alienation. Professionals in general enjoy an advantage despite the fact that the highly competitive nature of their jobs often mitigates against friendly, open peer interaction, and creates colleague-to-colleague alienation and even animosity.

Even fairly prestigious fields have rules and status distinctions that set limits to the individual's time, space, and energy. A professor who enjoys teaching and is well regarded by her or his students may find that she or he must publish, even if the effort required for publication takes energy away from teaching. A doctor may have to learn deferential behavior in relation to the chief of staff of the hospital. She or he will also learn to pick up cues about her or his own position at work—whether she or he is greeted by superiors in the hallway, and where her or his office is placed and how large it is.

In work, as in every other facet of life, communication comes through the allocation of space and through the words, gestures, and intonations of the people around us. If we seek to develop into a society of responsible and able adults, it is clear that the messages we receive at the workplace need to be altered. In the society at large, the concept of work as a potential means to self-actualization rather than as drudgery and alienation must be encouraged.

Learning with and from Peers

While much research and writing has been done on the importance of peer groups in adolescence (see Chapter 6), it is important to remember that peer contact and social learning from peers begins between siblings or in nursery schools, or as soon as the child realizes there are other kids in the neighborhood. Children begin to form peer groups as soon as they begin to make friends, and these friends and the activities in which they participate together are rich in social learning. One can imagine that to a young child the values, the interaction patterns, religious practices, and mealtimes of her or his own family must appear at first as universals. For children, time spent with playmates and visits to friends' homes are cross-cultural experiences in the sense that they provide a look at different allocation of time, space, energy, and money from that seen at home. Such real-life experience is likely to have a greater impact than influences such as TV and books, which also compete with the family.

In addition to sharing their home lives, children share their own world of play.[46] Some rhymes and oaths of friendship, for instance, have been passed on essentially unchanged for more than one hundred years by way of an oral tradition that has no firm curriculum and is removed from adult or professional intervention. Such a tradition is a rich source of learning. Children guide each other in the proper observance of special holidays, in the rites of secret clubs and blood brotherhood or sisterhood, how to cure warts, cleanse the spirit, levitate one another, and deal with loose teeth. There are jingles about teachers, parents, and policemen. Explanations of war, sexuality, drunkenness, and general success or failure are transmitted through jokes and jump-rope jingles.

As amazing as the stability of this lore and tradition may be, it is certainly matched by the incredible speed with which new verses, pranks, and parodies make the rounds. Iona and Peter Opie document a 1956 example of an alteration in a verse of "The Ballad of Davy Crockett" that turned up with "miraculous" speed in many counties of England and parts of Australia within a few months after the original ballad was released.[47]

Games and sports are another aspect of children's play and social learning. G.H. Mead has theorized that games are a form of role rehearsal. This thought has been echoed by Brian Sutton-Smith who has done substantial observation of children's play and informal learning. Sutton-Smith has said that play allows the child a chance to express both a consolidation of experience and an anticipation of what is to come.[48] It is easy to see that in many games children are trying on adult behavior, or anticipating a smaller step—as in dating games with Barbie and Ken dolls or imitation of the latest TV hero.

Games tend, of course, to fit the culture in which they are born and for which they constitute a form of preparation. They frequently include variations on the theme of pursuit and escape, attack and defense, and accumulation and deprivation or loss. Sutton-Smith has noted that no culture has only aggressive games with teams and winners or losers. Every culture has some cooperative games. Some forms of play hold order and disorder in tension; other games, like Redlight-Greenlight, include either validated cheating or simulated naughtiness. Depending on how children play it out, such opportunities may allow a child to be the authority, to defy authority, or simply to review and adjust to the order to things as they are. By opening their home lives to one another through their play, and in the direct sharing of information and values in simple conversation, young children affect the social learning of their peers.

As children grow into adolesence, peer interaction becomes even more salient. Because another chapter in this book will deal particularly with adolescent peer contact, however, we will move directly from the preceding material on childhood to a discussion of adulthood.

We should perhaps begin this consideration of adult peers by distinguishing between a "peer" and a "peer group." On the one hand, the term "peer" implies mutuality, and is defined as anyone with whom we share an equal potential for learning and teaching. We interact with different groups of peers in different environments—at work, church, school, on vacation, and in social or political clubs. In each context, our peers correspond to different facets of our lives.

Our "peer group," on the other hand, usually refers to the people we choose as our friends. The data indicate that most people form friendships with people who they feel are most like themselves. Close friends tend to have comparable incomes, levels of education, basic values, and models of reasoning. We choose our friends because they agree with us, are like us, or in some way have proven mutual compatibility. Some level of conformity (and this level may vary) is thus expected and enjoyed as harmony, predictability, and validation of our own lives. This is true even when the hallmark of the reference group is its nonconformity with the dominant culture, as in a closely formed gay community.

Whatever part our peer groups may play in restricting individual exploration and change, they also add much that is beautiful to our lives. They can provide comfort, by protecting the individual or the nuclear family from the worst aspects of isolation. They can provide a circle of warmth that fosters a sense of security and a sense of collectivity that may come as an antidote in a highly competitive culture. They provide a relatively safe and stable social environment into which the individual can incorporate new learning without feeling the entire world is turning topsy-turvy. In fact, the data on the diffusion of innovation indicates that potent change agents invariably share many of the patterns and values—that is much of the culture—of the people whom they affect.[49] With this in mind, to the extent that we identify with them, mass media models can also function as close friends, in the broader sense, and as members of the peer group.

Among peers who are not chosen friends—those who come with the job, the neighborhood, or the extended family—the conformity that is expected may be grating rather than comforting. Such groups may restrict individual experimentation and new social learning by withholding support for new and unexpected views and behavior. We would posit that the extent of this restriction varies between social classes as a function of geographical and occupational mobility.

Because mobility brings the opportunity both for relative anonymity and for meeting new friends, it is conducive to social learning. It lets us manifest and test latent learning by affording an opportunity to try out new behavior and opinions, since no clear picture of "what is expected" yet exists. Mobility also puts us in touch with new people and thus with new values and ideas and behavior patterns. The relative anonymity of being apart from people who "knew you when" is one facet of privacy and is conducive to exploratory change much like the privacy of one's own room, or children's pretend games out of sight of their parents.

Lest we go overboard in glorifying transiency, we should reiterate an earlier point. People do not automatically try out every new idea that they encounter from any source. A potential source of new learning must affect us with great force in order for it to influence or change the learning that we first encountered in the family. And no learning can take place unless we are open to the message transmitted from this new source.

In summary then, peers and peer groups are a significant source of social learning during all phases of life, except infancy. Peers learn from one another through shared experiences, verbal and nonverbal interaction, and observation. Like family members, peers are people with whom we negotiate in what Kantor and Lehr have described as the individual quest for affect, power, and meaning in our lives.[50] With peers, we both repeat and alter the routes to these goals that have been established in interaction with our families. While peers are often given to us by circumstances, our peer groups are formed through individual choice and are in many senses a reflection of who we are. In varying degrees, our peers provide both impetus for and restriction on change.

Learning from Mass Media. The power of the mass media both to inform and to influence people's attitudes and behavior has been much noticed. Before TV or radio, books like Sinclair Lewis's *The Jungle* or Harriet Beecher Stowe's *Uncle Tom's Cabin* were credited with bringing to light and speeding the rectification of severe social ills. Television has been credited, rightly or wrongly, with increasing the general level of violence in our society, and even with inspiring specific antisocial acts such as hijackings. Whatever the result of future research on TV and violence, it is clear that the effects of the mass media are varied, complex, and far-reaching. In this section we will look at some of the research that has been done on the mass media as a source of social learning. As we focus on the allocation of resources, we will take special note of recent research that has studied the way in which viewers use television.

Before we go any further into the mass media's effects, we should remind ourselves of the many forms of communication included in the term"mass media." Most research in the field includes only television, radio, newspapers, books, magazines, and comic books. The recording industry also deserves inclusion in this list, although it clearly overlaps with radio and television. In the last few years the public has been spending more on records and tapes than on other forms of entertainment. The recording industry's revenue for 1978 ($4.2 billion) was more than that of either the movie industry or all sporting events; the year's bestselling album, *Saturday Night Fever*, sold over 15 million copies in the United States (one for every five households).[51] For the purposes of our own study, the particular significance of these figures is that the record industry is aimed primarily at a young market, and is a large source of information about sexuality. The Reverend Jesse Jackson recently expressed concern about the effects of the music industry on the values of youth, particularly black youth. The power of this industry comes from at least three sources: the creation of cult figures such as Bob Dylan and the Beatles, the content of the lyrics, and the music itself. We are all aware of the effect music has on our moods, and we choose music to enhance a current mood, or in order to move us to another mood. The music that we do not control also has a profound effect on us. Office and supermarket managers have known for years that the music (or Musak) that they play in their spaces will affect how much people work or buy.

In recent years, television has been the most intensively researched of the mass media. In some cases it may be justifiable to use the findings of television research to hypothesize about other media. On the other hand, real differences distinguish the media themselves as well as the social situations in which they are used. For instance, because both television and film provide a sensory experience that often leaves less to the imagination than printed media do, one might assume that their effects would be quite similar. Yet if one compares TV and film in terms of their typical settings, the rules regarding viewing, and the cost to the viewer, one soon realizes that there will be differences.

If television is the most studied medium, most television research thus far has concentrated on the relationship between TV content and viewer attitudes—especially in the areas of violence, gender-role stereotyping, the depiction of minorities, and vocational portrayals.[52] Another chapter in this volume will examine the literature on TV programming and its effects in some detail (see Chapter 4). Here, just two brief points will be made as regards content. First, the bulk of published research came out before the spate of recent programs emphasizing sexual issues or problems. What effect this recent phenomenon will have on American life has yet to be researched. Second, TV research thus far appears to raise an important policy issue. To what extent do we want television to accurately reflect social reality or indoctrinate us in the dominant beliefs of this society; and to what extent would we have television improve upon the real world (by being less sexist, racist, or violent) in the hope that it may move us in that direction.

Parents have been concerned for years about the influence of television on their children. They have worried about children being exposed to violence and explicit lovemaking. Television often presents a style of life and a set of values that differ drastically from those of ordinary home life. For the most part, television—an industry controlled by big business—has seemed beyond parental control. A parent can turn the tube off, but cannot influence program policy. Consumer groups like Action for Children's Television are an attempt to affect television programming, thus reclaiming some control of the socialization process. The local programming potential of cable television may provide another avenue of parent or community control.

Aside from the content of television programs and that of other media, additional factors must be considered in order to fully understand and be able to predict the content's effects. Elihn Katz, Jay Blumler, and Michael Gurevitch, for instance, have argued that mass media research must involve (and evolve) a theory of the role of media in people's lives. One must examine the motivations for viewing, reading, and listening. They call this a "uses and gratifications" approach to media research and suggest that

> such an "approach has proposed concepts and presented evidence that are more likely to explain the media behavior of individuals more powerfully than the more remote sociological, demographic and personality variables." (Weiss 1971.) Compared with classical effects studies, the uses and gratifications approach takes the media consumer rather than the media message as its starting point. . . . Thus, it does not assume a direct relationship between messages and effects, but postulates instead that members of the audience put messages to use, and that such usages act as intervening variables in the process of effect.[53]

This theory implies, for example, that the simple breakdown of media material into advertising, news and public affairs, and entertainment does not hold. Studies have shown that some people attend to the news for essentially escapist reasons. For others, the so-called Perceived Reality of Television (PRTV) on non-news shows is so high that it may serve a newslike function in describing reality (however accurately or inaccurately).[54] People may watch the same TV show for different reasons, and—guided by their own interests and concerns—they may tune in to different features of the show.

In recent years, PRTV has stirred some interesting research. Bradley Greenberg and Byron Reeves have said that, "to the extent that television content is the sole or principal source of information for a child on some particular issue or social situation, that information is more likely to be judged by the child as realistic and to be accepted." The next step is that the more TV is perceived as real, the more likely it is to be modeled. In general, PRTV decreases with age. Girls have a slightly higher PRTV than boys. Socioeconomic status and PRTV Correlate inversely. Blacks tend toward a higher PRTV than whites even when socioeconomic status is controlled. Gerson and others have argued that mass media are more potent in this way for blacks than for whites because their power is increased when other socializing agents like home and school are in conflict.

Greenberg and Reeves have looked at the effect of interpersonal communication on children's PRTV, and have found that "interpersonal influence plays an extremely prominent role in the child's assessment of the reality of television. It does so to some extent for content areas in general; however, its chief function is in aiding the child to define the reality of specific television roles."[55]

An obvious suggestion of both the "uses and gratifications" theory and PRTV might be to look at TV viewing in the context of our theory about the allocation of resources. The effect of TV on children's social learning (or on how real children believe its messages to be) will depend to some extent on the children's perception of how seriously their parents or their peers appear to regard it. Is it accorded large amounts of time, space, energy, and money in the home or by close friends? In this respect, it is interesting to note that 45 percent of American homes now have the TV on during dinner—traditionally a time of family talk.[56] The statistics mentioned earlier on phonograph record sales are also relevant. What do these figures mean in terms of teenagers listening to song after song about love affairs? How real are these stories to people who may be experiencing their first confusing encounters with love?

Another aspect of the foregoing theories might be that gratification comes for a variety of reasons. "Studies have shown that audience gratifications can be derived from at least three distinct sources: media content, exposure to the media per se, and the social context that typifies the situation of exposure to different media."[57] When one reads a Nero Wolfe mystery, for instance, she or he may be folfowing the adventure, learning from Wolfe's recipes, enjoying quiet time alone, or perhaps sharing the story with other members of the family.

The effects of the mass media are multitudinous and are likely to stimulate us on many levels. The effects are also hard to isolate and document in a rigorous scientific manner. In much the same way that families reveal their priorities through the allocation of time and space, the mass media do this with air time and column space. They tell us who and what is important to know and care about, and they control much of what we know about these people and issues. Over 95 percent of the homes in this country have TV sets; 98 percent have radios. The mass media therefore have an incredible homogenizing potential, mitigated only by selectivity in their use. This selectivity, of course, is subject to the limits of available choice. (We have not discussed the mass media as commercial industries. Their commercial nature puts some constraints on what can or cannot be learned from the media.) Finally, of course, the mass media provide more role models, both in the fictional characters they portray and in the "superstars" they create. These may be most effective when the media stories give us access to aspects of life that we do not share in our peer groups.

Learning from Experts

In a society as highly specialized as our own, the role of experts in an individual's social learning is very complex. One problem is to distinguish the experts from the pseudoexperts, the expertise from the mythology. The advertising industry is constantly referring to "a doctor's study" or using pseudoscientific language and charts to sell products. For the most part, however, this is probably a reflection of the extent to which this country respects and relies on experts and authorities. We need not credit Madison Avenue with creating the mentality, only with exploiting it and thus perpetuating it.

Ivan Illich has said that above all else what an individual learns from experts, when expertise is equated with professionalism, is self-mistrust. Illich sees compulsory formal education and the deification of diplomas as the core of the problem.

> Neither in North America or in Latin America do the poor get equality from obligatory schools. But in both places the mere existence of school discourages and disables the poor from taking control of their learning. Once we have learned to need school, all our activities tend to take the shape of client relationships to other specialized institutions. Once the self-taught man or woman has been discredited, all non-professional activity is rendered suspect.[58]

There are at least three types of contact that an individual may have with experts. (1) One may have a face-to-face encounter with a particular teacher, doctor, social worker, or auto mechanic. (2) One may read or hear messages from particular experts through the various mass media, including word-of-mouth. (3) Through the influence of their professional organizations, lobbies, and government work, experts contribute heavily to the shaping of our environment and our culture.

In this section we will explore each of these types of contact. In particular, we will look at these contacts as they affect the sense of power in order to better understand Illich's assertion that we learn to mistrust our own experience. We must bear in mind that encounters with experts differ according to class, race, and sex. These characteristics will contribute to both the expert's and the client's sense of the extent to which they are in an exchange between peers or nonpeers. Since very little research has been done specifically on the role of experts, much of this discussion will be an application of points made earlier in the paper to the particulars of a client-expert relationship.

As in any encounter, many of the realities of a face-to-face encounter with an expert are perceived through the expenditure of time, energy, and money. Typically, the client must use his or her time and energy to make the encounter possible and in the end pays the professional. All of this is a reflection of the fact (or the mutual belief) that the expert has something that the client needs. This fact is, in turn, an aspect of power. If clients are late, they may well lose their appointments. If experts are late, they are awaited. In many clinics and welfare offices, all the clients for a given morning or day may well be scheduled A.M. and left to wait indefinitely.

Once the client and the expert are together, variations in the exchange obviously occur. Every such encounter, however, is based on the fact that that professional and his or her colleagues have an effective monopoly on something that could benefit the client. During the encounter the professional controls how much he or she shares. The number of questions permitted, the length of the encounter, and the extent to which the client is able to feel at ease all affect this information-sharing. Thus, experts may or may not tell the client what they are doing and why—why, for instance, they are tapping our knees with little hammers or why they are asking us how we did in elementary school twenty-five years ago.

The language of expert to client may be in a coded jargon, which more than anything else communicates distance. Through persistence the client may be able to affect these exchanges, but we must remember two things. First of all, the professional still controls whether or not he will let that happen. Secondly, whether or not the client even presumes to do that will vary according to prior training in expert-client relationships as well as according to race, sex, and status. This is not to say the client will learn nothing about law or medicine or auto mechanics. But what is learned is in the expert's control and will probably be limited.

This process can be understood in terms of nonformal, informal, and incidental learning. In the situations described above, most of the learning is incidental or informal. That is, the expert-client encounter is rarely one in which both parties are present for educational purposes. Certainly, neither would say that they are there so that the client may learn deference in relation to professionals and a mistrust of self. Typically, one party or another is present either with the intent to teach or with the intent to learn. For instance, a doctor may want to teach a patient how to take better care of her or himself, while the patient wants only to be taken better care of by the doctor.

Our hope would be that increasingly both parties will view these encounters as potential learning experiences, and thus they will move into the realm of nonformal learning situations. To some extent, this is already happening. The increased use of paraprofessionals has often meant that clients get more attention and more education from social service agencies. Clinic waiting rooms often have leaflets and videotapes that explain causes and prevention of common complaints. The women's movement and various consumer groups have encouraged clients to ask questions and to demand that professionals and agencies be accountable to their clientele.

In the second type of encounter—the one that takes place via the mass media—the expert is relatively inaccessible and sometimes absolutely unknown. One example might be Dr. Spock. Most of his audience has not met him face-to-face. Moreover, he is in no way accountable to us. We hear from him according to his schedule, and his messages come through the mass media—an arrangement that absolutely prevents dialogue. The audience has no power, no ability even to ask questions. We hear from the media and friends that he is considered the leading expert on baby rearing, and we put that together with all the messages we have received about experts in general.

The basis of our respect for such experts may be confused because we are often informed about aspects of these people other than their expertise. Through vehicles like *People* magazine and "The Today Show," we may end up knowing as much (or more) about an expert's childhood, hobbies, or marital relationship as we do about his or her character, work habits, or even credentials. Such magazines and TV shows also constantly need fresh entertainment, different heroes, and consequently new experts. In a mass media-consumer culture, experts become stars. And as stars, they rise and fall. The media, like a spotlight, directs our attention first to Expert A, then to Expert B, and so on.

Experts, such as Dr. Spock, also affect us indirectly in their influence on more local experts with whom we have face-to-face encounters. Thus, Dr. Spock has affected many Americans, both because he published a book in lay language and has published in popular magazines, and because he has been a source of learning for hundreds of thousands of pediatricians.

Finally, let us consider the social learning that comes from all those unseen experts who plan highways through or around our neighborhoods, spend our money on wars or schools of midwifery, design buildings with windows that don't open, invent aerosol cans that destroy the ozone layer, or develop automobile seatbelts that truly save lives. These people have an impact on the national or international use of time, space, money, and every form of energy. They occasionally determine our social learning in areas that may range far beyond their fields of expertise. The men who designed and supported the designing of the atomic bomb, for instance, did not teach us much about bomb-building or nuclear physics, but they have definitely altered our environment and thus affected our social learning. Hannah Arendt has said of the generation of people who came of age under the shadow of the bomb that if you ask these people what the world will be like in ten years, or what they would like to be doing in five years, they frequently preface their responses with "provided there is a world in five years" or "if I am still around in five years."[59] Such a response is a manifestation of a particular chunk of social learning, a core attitude toward life, that was born of particular circumstances, but which continues to outlive those circumstances. We can assume that it was unintended, unpredicted, and unwanted by those experts who contributed to shaping the environment that fostered it.

In summary, we can say that social learning from experts includes, but is not limited to, their field of expertise. What we learn about their field of expertise is largely in their control. They exercise this control through the amount of time they share with us, through how comfortable they help us feel, through what they choose to tell us, and through the energy they expend in communicating to us in our own language. Other social learning about ourselves and our world comes from these experts both in how they treat us in personal encounters and in the impact that they have on the world in which we live.

Learning from Religion

The influence of religion is pervasive and touches every member of society, whether that member considers him or herself to be religious or not. Our legal codes have their origin in religious doctrines; our national holidays originate in the concept of holy days set off as special times of the year. In court procedure, we are made to swear on the Bible as a means of enforcing the truth of our words.

In discussing religion and the church as a social learning environment, we are not primarily concerned with the formal aspects of religious learning—doctrine, dogma, and theology. Instead we are concerned with the effects of religious teachings as they are experienced in rituals and practices, through symbols, relics, and architecture, and through the implicit regulation of behavior through the management of the life of church members.

Rituals and Religious Practices

Rituals and religious practices include celibacy, confession, sermons and worship, membership rituals, baptism, marriage and death rituals, anticipation of salvation or damnation, and holidays. Ritualization helps to perpetuate the authority of the church and solemnizes much of its teachings. To the individual, ritualization communicates a special meaning about these activities. The formality and aloofness of many rituals communicate an aura of specialness or sanctity which may become "functionally autonomous." The time that is devoted to these rituals, the personal energy that is required in order to carry them out, and even the amount of space set aside for particular rituals all communicate messages about their importance. Whether the marriage ceremony takes minutes, hours, or a whole day; if there is no special ritual for divorce; comparative length and complexity of death rituals—all of these communicate messages to the individual about the meaning of these acts in his or her life.

Rather than underlining the significance of an act, ritualization may sometimes drain it of its meaning. For example, the traditional ritual of exchanging marriage vows may communicate little about the feelings and commitments of people to each other; or confession may become an "act of behavior" rather than an act of belief, or holidays, such as Christmas, become expressions of something quite different from their original meanings.

On the other hand, rituals sometimes maintain original meaning but expand on it to meet current needs and thereby communicate originally unintended messages. For example, in Judaism two of the major holidays, Pesach and Chanukah, are celebrations of freedom. Pesach celebrates the fight for freedom from physical enslavement in Egypt and Chanukah celebrates the struggle for religious freedom under Greek rule. Both of these holidays impart the explicit message that freedom is important enough to fight for (and sometimes suffer for). Both of these holidays celebrate a liberation that ultimately was the result of a "miracle," conveying, perhaps, an explicit message that military and political organization are less important than righteousness and faith. The annual retelling of these stories stresses the importance of history, of not forgetting, of a sense of continuity with tradition, and of belonging. The importance of belonging is further supported by the fact that the celebration is with the entire community at the temple and with the extended family and friends at home. Of course, exactly what is learned from such rituals depends on how they are handled by individual rabbis and parents. From the same holiday story, some people may develop a deep compassion for the oppressed of this world, while others may develop a nationalistic identity as a Jew.

Architecture, Symbols, and Relics

Some of the power, awe, and mystery of religion comes from the structure of churches and temples. The high ceilings and the ornamentation lend a sense of grandeur which one may share for awhile, or which may make one feel small and insignificant. In the Jewish religion, the holy scrolls are usually hidden beneath layers of jeweled velvet behind a series of locked doors and cabinets. To many this means the scrolls are important, but it also means they are inaccessible. The knowledge and the enlightenment that they contain seem like secrets.

In sharp contrast to this Judaic grandeur, is the simplicity of a Quaker meeting house. The building embodies a message found in many Quaker hymns and writings emphasizing the beauty and serenity that come with simplicity. This simplicity is also exemplified in the music and administrative structure of the church. As in some other Protestant religions, this simplicity manifests the belief that one ought to have a direct relationship with God.

The church building is also a symbol in the society as a whole. Until the later half of this century, it was the tradition for the church to be one of the first buildings constructed when a new community was established. Today, many communities are built around shopping centers or factories. The time, money, and energy given to the construction of the church building contains messages about how the church fits into the rest of life.

Other symbols and relics also communicate messages. For example, the cross and the statues in the Catholic church are important reminders of Christ "who died for the sins of the world" and the goodness of people inspired by God, exemplified by the saints and the Virgin Mary. Relics or ancient religious artifacts, sites, and buildings can also become very important sources of religious messages and often evolve into a driving motivation for a particular religious belief or activity. Witness the attempt of the Crusades to free the city of Jerusalem from the so-called infidels.

Certain clothing can also be a strong symbol. Religious clothing—such as the nun's habit, the priest's collar, or the Jew's yarmulke—symbolize an entire way of life and serve to distinguish one as special and distinct. The clothing carries with it the requirement of certain behavior, and predisposes attitudes toward the wearer.

Management of Life of Members

Perhaps the most important means of learning and communicating through religion and churches is in the management of most of a person's life. Religion is with a church member whether she or he is in church, or practicing a religious rite, or not. The attitudes, beliefs, and behavior instilled by the church influence work, recreation, and family and other interpersonal relations. This does vary, of course, depending on the devoutness of belief or the church affiliation.

Part of this management is made evident in each religion's general approach and in specific sanctions regarding the responsibility and authority of church members and officers. In Catholicism, for instance, power and knowledge reside in church officials. Such authority has been transmitted through Christ to Peter to the disciples to one's own parish priest.

In some churches there is also a tendency to manage much of the social as well as the religious life of its members. Messages are communicated about how much time you should give to the church, and also about how you should deal with your own private space. In some Protestant sects, for example, you are expected to make your home available to others whenever the church asks.

A further religious influence comes from the effect of the Judeo-Christian ethic on other institutions in our society. This is partly a matter of ancient history and heritage, but it is also partly current history. The historical point is illustrated by the extent to which such things as the Commandments or the Code of Hammurabi have shaped our current legal and penal institutions. The current point is illustrated by the many forays that religious groups make into the public arena. Martin Luther King's power came partly from his connection to the church. That power consisted of money, people, and a potent sense of righteousness. The Catholic church has brought money, organization, discipline, and moral authority to bear in the controversy over abortion.

The degree to which the church becomes involved in social and political causes carries many complex messages that echo throughout the life of the individual, as well as throughout the society as a whole. Important messages are communicated if the church building contains posters about war, hunger, or human rights struggles, and if sermons and lectures are presented in the church on these topics.

Clearly, then, religion is a strong source of social learning. The channels of communication are numerous—including the specific doctrine, the forms of rituals, the physical setting, the authority structure, the social contact involved, and the host of social and political attitudes and traditions within each religious community. It has been said frequently that many Americans restrict their religious participation to church attendance on Sunday morning. It probably can be safely said that even if attendance at church is thus limited, religious practice in this society has imparted a consciousness and a set of values and attitudes much of which is carried with us always and much of which is reinforced through other social channels in our society.

CONCLUSIONS AND WORK TO BE DONE

At this point, we will mention two pitfalls of examining social learning in the segmented or compartmentalized manner of this chapter. First, people do not experience their lives and their learning as being chopped up into separate pieces. The formation of values, attitudes, and behavior patterns does not result from the simple addition of keenly noticed moments of learning. Rather, social learning is the absorption and distillation of many noticed, half-noticed, and unnoticed messages that come at us from every direction. Some messages emanate from multiple sources and are repeated in a variety of ways in numerous situations; others are relatively infrequent. In the section on mass media, for example, we say that we need not concern ourselves so much with any single example of television violence or any single protrayal of blue-collar workers, but rather with the repetitious pattern of such portrayals and the lack of alternative social messages. This principle applies as well to the examination of the various social learning environments. Are people being encouraged to abandon their potential for creative and independent thinking only at work, or in other realms of their lives as well? Is the view of women as inferior human beings put forth only by our religious institutions, or is it embodied in and promulgated by other social institutions?

The message patterns are difficult to recognize because such relationships constitute the very structure of the environment in which we live and learn. The easiest language to learn is the one that we hear every day. The easiest imaginable arrangements of work life and home life are the ones that we experience all the time. We absorb certain assumptions about life without even knowing it—by living in conditions based on those assumptions. It is far more difficult to imagine in any detail a world, a way of life, a form of family, or experience of sexuality that we have never known. And, indeed, even when we can imagine them, we may learn that they are impossible to realize within our given conditions.

The second problem with looking only at individual sites of learning is that it is hard to recognize all of them. Certainly, the ones we have mentioned here do not account for the whole of any individual's life. If we envision a person's life as a two-dimensional plane, and we take all the incidents of social learning mentioned in this paper—plus all the other such moments that we can think of—and plot them as points on that plane, there will still be a lot of space between those points. It is not empty space; it is space filled with potential for social learning. Our task is to recognize this social learning even when it does not fall within an easily defined learning environment. We can help ourselves in this process if we try to notice not only what is present, but what is absent as well. The vitamins that are absent from our diets are important; such vitamins affect our health, sense of well-being, and energy level, as well as our moods. The trees and open spaces that are missing from our cityscapes are important. Their absence marks the absence of particular experiences and a certain relationship to nature and time in our lives.

Similarly, wide streets and narrow sidewalks declare that America drives; it does not walk. Promenading is not an American pastime. Nor do we have many sidewalk cafes in which to enjoy a leisurely cup of coffee or a drink and conversation or crowd-watching. These facts are a reflection of and a factor in creating or perpetuating the American citydweller's attitude toward time and outdoor space. Most people who are out of doors in a city are just getting from one place to another; such an attitude makes that time and space "lost." Such examples merely suggest the kind of absent resources that must be recognized in order to fully visualize the many aspects of social learning.

With those cautions about what may be accidentally omitted through this approach to social learning, let us now recall what has purposefully been defined as outside the scope of this chapter, and consider the work that has yet to be done.

Each of the learning environments that have been briefly presented here must be examined in depth. This examination must include historical analysis of the development of our contemporary social institutions. We must look particularly at the ideologies and conditions that fostered the evolution of our current forms of family, religion, and work. Which aspects of these forms are still appropriate to the lives that we lead and want to lead, and which aspects call for change?

Having some insight into the locations and mechanisms of social learning, we must explore the content of this learning. What are the messages that we are sending and receiving? This has been largely ignored in this chapter and must be addressed in a thorough and systematic manner. The data from a recent study in Cleveland will begin to answer some of our questions about sexual learning in the home.[60] We must find ways to pose and answer questions about social, and particularly sexual, learning in every environment.

As we construct this picture of social institutions, we must pay special attention to their interrelationships. How do these institutions influence each other and then jointly touch each of us? How do they enhance or impede each other's effects? For instance, we say that the family is of supreme importance. Parents are told by experts, friends, and the media that they should spend more time with their children; Freudian theory, still strong in our culture, emphasizes the importance of the early years. Yet maternity and paternity leaves from work are short or nonexistent. We do not support people to tend their children. School schedules and work schedules vary so that parents and children often are not free at the same times. In contrast, for instance, many kibbutzim have extended lunch breaks that parents and children often spend together.

These social conditions—these uses of public time, space, energy, and money—and these demands made on an individual's resources must be considered in formulating social policy that relates to sexuality (or to any other realm of human experience). If we want to see changes in these areas, then we must make those changes possible by creating the structural and material conditions that encourage them.

If we do not like current ideas about gender identity, family roles, intimacy, beauty and ugliness, or acceptable sexual expression, then we must discover how they are being learned. If we want to live in a more androgynous world and be more comfortable in our bodies, if we want to be less afraid of intimacy and less ambivalent about sexual expression, then we must discover and create social conditions that will allow for and reinforce the desired attitudes and behavior. That is the task that lies before us.

NOTES TO CHAPTER 2

1. John Gagnon, *Human Sexualities* (Glenview, Ill.: Scott, Foresman and Co., 1977), p. 2.

2. When we speak of social learning we are referring principally to what might be called the Albert Bandura school of social learning. Clearly, however, we, and not Albert Bandura, are responsible for all statements in this paper, unless otherwise indicated.

3. The following is an overview of some theory and research in the field of social learning. As such, it is not intended to be an exhaustive review of the literature or a state-of-the-art appraisal. It is intended to be a review of some of the main points we think are important in understanding sexual learning as a social learning phenomena.

4. David Krech, Richard S. Crutchfield, and Norman Livson, *Elements of Psychology* (New York: Knopf, 1969).

5. H.H. Toch and R. Schulte, "Readiness to Perceive Violence as a Result of Police Training," *British Journal of Psychology* 52 (1961): 389–394.

6. A. Bandura and Aletha C. Huston, "Identification as a Process of Incidental Learning," *Journal of Abnormal and Social Psychology* 63 (1961): 311–318; B.A. Henker, "The Effect of Adult Model Relationships on Children's Plan and Task Imitation," *Dissertation Abstracts* 24 (1964): 4797; P.H. Mussen and A.L. Parker, "Mother-nurturance and Girls' Incidental Imitative Learning," *Journal of Personality and Social Psychology* 2 (1965): 94–97.

7. A. Bandura and Carol J. Kupers, "Transmission of Patterns of Self-Reinforcement through Modeling," *Journal of Abnormal and Social Psychology* 60 (1964): 1–9; A. Bandura, D. Ross, and S.A. Ross, "Imitation of Film-Mediated Aggressive Models," *Journal of Abnormal and Social Psychology* 66 (1963): 3–11.

8. A. Bandura and F.J. McDonald, "The Influence of Social Reinforcement and the Behavior of Models in Shaping Children's Moral Judgments," *Journal of Abnormal and Social Psychology* 67 (1963): 274–281.

9. A. Bandura and W. Mischel, "The Influence of Models in Modifying Delay of Gratification Patterns," *Journal of Personality and Social Psychology* 2 (1965): 698–705.

10. R.R. Sears, E.E. Maccoby, and H. Leving, *Patterns of Childrearing* (Evanston, Ill.: Row, Peterson, 1957); A. Bandura and R.H. Walters, "Aggression," in *Child Psychology: The Sixty-Second Yearbook of the National Society for the Study of Education, Part I* (Chicago: The National Society for the Study of Education, 1963), pp. 364–415.

11; W. Mischel, "A Social Learning View of Sex Differences in Behavior," in *The Development of Sex Differences* ed. Eleanor Maccoby (Stanford: Stanford University Press, 1966), pp. 56–81.

12. W. Mischel and R.M. Liebert, "Effects of Discrepancies between Observed and Imposed Reward Criteria on Their Acquisition and Transmission," *Journal of Personality and Social Psychology* 3 (1966): 45–53.

13. A. Bandura, *Principles of Behavior Modification* (New York: Holt, Rinehart, and Winston, 1969), p. 19.

14. Benjamin Whorf, "Science and Linguistics," in *Readings in Social Psychology*, eds. Eleanor Maccoby, Theodore Newcomb, and Eugene Hartley (New York: Holt, Rinehart, and Winston, 1958), p. 5.

15. George Orwell, "Politics and the English Language," *Inside the Whale and Other Essays* (New York: Penguin Books, 1957), pp. 143–157.

16. John B. Carroll and Joseph B. Casagrande, "The Function of Language Classifications in Behavior," in *Readings in Social Psychology*, eds. Eleanor Maccoby, Theodore Newcomb, and Eugene Hartley (New York: Holt, Rinehart, and Winston, 1958), pp. 18–51.

17. Ibid., p. 22.

18. Robin Lakoff, *Language in Society* (New York: Harper & Row, 1975), cited in Nancy Henley, "Power, Sex, and Nonverbal Communication," *Journal of Sociology* 18 (1973): 1–26.

19. Ibid., p. 6.

20. Maija S. Blauberg, "On 'the Nurse Was a Doctor'" (paper presented at the Southeast Conference on Linguistics, 1975).

21. J. Sachs, "Development of Oral Language Abilities from Infancy to College," (Final Report, Grant #OEG–0–9–160440–4144(010), U.S., Department of Health, Education and Welfare, Office of Education) 1972.

22. Dan Slobin, Stephen Miller, and Lyman Porter, "Forms of Address and Social Relations in a Business Organization," *Journal of Personality and Social Psychology* 8 (1968): 289–293. In Nancy Henley, "Power, Sex, and Nonverbal Communication," *Journal of Sociology* 18 (1973): 8.

23. Erving Goffman, "The Nature of Deference and Demeanor," *American Anthropologist* 58 (1956): 473–502. In Erving Goffman, *Interaction Ritual* (New York: Anchor 1967), pp. 47-95. In Nancy Henley, "Power, Sex, and Nonverbal Communication," *Journal of Sociology* 18 (1973): 7.

24. Nancy Henley, "Power, Sex, and Nonverbal Communication," *Journal of Sociology* 18 (1973): 3.

25. Michael Argyle, V. Salter, H. Nicholson, M. Williams and P. Burgess, "The Communication of Inferior and Superior Attitudes by Verbal and Nonverbal Signals," *British Journal of Social and Clinical Psychology* 9 (1970): 222–231. In Nancy Henley, "Power, Sex, and Nonverbal Communication," *Journal of Sociology* 18 (1973): 3.

26. Edward T. Hall, *The Silent Language* (New York: Doubleday and Co., 1973), p. 15.

27. *Metropolis: Values in Conflict* Svend Reimer, ed. (Belmont, CA: Wadsworth Publishing Company, 1964).

28. All references to Dr. Chester Pierce come from private conversations and from lectures delivered for his course "The Roots of Violence," Harvard Graduate School of Education, 1976.

29. *The American Heritage Dictionary of the English Language* (New York: Dell Publishing Co., 1973), p. 239.

30. Urie Bronfenbrenner, "Experimental Human Ecology," cited in James Garbarino and Susan Turner, *Television and Vocational Socialization* (Cambridge, Mass.: Marketing Science Institute, 1975).

31. Eli Zaretsky, "Capitalism, the Family, and Personal Life: Part I," *Socialist Revolution* 3, 1 and 2 (January-April 1973): 83. Zaretsky's work has greatly influenced our understanding of the development of the contemporary family and its relationship to other social institutions.

32. Eli Zaretsky, "Capitalism, the Family, and Personal Life: Part II," *Socialist Revolution* 3, 3 (May-June 1973): 84.

33. David Kantor and William Lehr, *Inside the Family* (San Francisco: Jossey-Bass, 1975), p. 37.

34. Maxine Wolfe, Mary Schearer, and Robert S. Laufer, "Private Places: The Concept of Privacy in Childhood and Adolescence," (paper presented at Environmental Design Research Association Meetings, Vancouver, 1976), p. 30.

35. Kantor and Lehr, *Inside the Family*.

36. S.H. Chaffee and J.M. McLeod, "Coordination and the Structure of Family Communication," cited in Jane D. Brown, "The Role of Communication in the Development of a Sex Role Orientation," (paper presented to the Theory and Methodology Division Association for Education in Journalism, San Diego, 1974), p. 14.

37. Kantor and Lehr, *Inside the Family*.

38. E.P. Thompson, *The Making of the English Working Class* (New York: Vintage Books, 1963).

39. Barbara Garson, *All the Livelong Day: The Meaning and Demeaning of Work* (New York: Penguin Books, 1977).

40. Ibid., p. 95.

41. Ibid., p. 171.

42. Ibid., p. 88.

43. Ibid.

44. Christopher Argyris, *Personality and Organization: The Conflict between the System and the Individual* (New York: Harper & Row, 1957).

45. Jane Loevinger, *Measuring Ego Development* (San Francisco: Jossey-Bass, 1970).

46. Iona Opie and Peter Opie, *The Lore and Language of School Children* (London: Oxford University Press, 1959).

47. Ibid., pp. 7 and 118–120.

48. Brian Sutton-Smith, "Games, the Socialization of Conflict," (paper presented to the Scientific Congress, 1972).

49. Everett Rogers, *The Diffusion of Innovations* (New York: Free Press of Glencoe, 1972).

50. Kantor and Lehr, *Inside the Family*.

51. Peter W. Bernstein, "The Record Business: Rocking the Big-Money Beat," *Fortune*, 23 April 1979, p. 59.

52. For a good summary, see James Garbarino and Susan Turner, *Television and Vocational Socialization* (Cambridge, Mass.: Marketing Science Institute, 1975).

53. Elihu Katz, Jay G. Blumler, and Michael Gurevitch, "Uses and Gratifications Research," *Public Opinion Quarterly* 37 (1973–74): 509–523.

54. Bradley S. Greenberg and Byron Reeves, "Children and the Perceived Reality of Television," Journal of Social Issues (1976): pp. 86–97, vol. 32, no. 4.

55. Ibid., p. 16.

56. James Hayden, "TV or Not TV?" *Bill Moyers' Journal*, 23 April 1979.

57. Katz, Blumler, and Gurevitch, "Uses and Gratifications."

58. Ivan Illich, *Deschooling Society* (New York: Harper & Row, 1971).

59. Hannah Arendt, *On Violence* (New York: Harcourt, Brace and Jovanovich, 1970).

60. Elizabeth Roberts, John Gagnon, and David Kline, *Family Life and Sexual Learning* (Cambridge, Mass.: Project on Human Sexual Development, 1978).

Chapter 3

Work, the Family and Children's Sexual Learning

*Cathy Stein Greenblat**

I. INTRODUCTION

Of all the institutions involved (or potentially involved) in children's sexual learning, the family is the one whose role is most generally acknowledged. The role of a parent or parents in teaching about nudity and modesty, genital differences, touching, affection and lovemaking, and reproduction is widely recognized and endorsed. Parents also serve as models for what it is appropriate for a male or female to do, to be, and to feel. Verbally and nonverbally, they give explicit and implicit gender messages that are highly influential in the child's early development. Finally, children first learn in the family about emotional closeness—how it is to be shown, as well as how likely and important it is.

The impact of the work environment on these elements of sexual learning are less obvious. In modern societies, extended years of schooling, child labor laws, and the belief that childhood is a special time in the life cycle mean that few children have serious work roles. Paper routes and baby-sitting jobs are usually not assumed until sometime after puberty. Moreover, since work and home are increasingly geographically separate, few children spend time in or even see their parents' places of work.

*Cathy Stein Greenblat is coauthor of the recent book *Life Designs: Individuals, Marriages, and Families*; a study of family formation and variation during the life cycle. Dr. Greenblat is Professor of Sociology at Douglass College, Rutgers University; a consultant to Little, Brown and Co.; and Associate Editor of *Teaching Sociology*.

Nonetheless, children are connected to the work world through their parents. Since there are powerful (though different) work norms for men and for women, all adults—whether or not they are employed—have important relationships to the work world. While the "bread-winning" father is the child's typical connection to this realm, there are several other possible points of connection. Among these alternatives are the increasing number of two-parent two-career families, families in which the mother is the sole breadwinner, or families in which neither parent is employed. Moreover, single-parent families raise an entirely different set of possible connections to work for the child.

In one way or another the child has at least a vicarious relationship with the work world, and the effects of the work world and the effects of the family on the child's sexual learning do not operate independently. While most of the effects of the work world are indirect, the work relationships of the parent(s) may conspicuously affect family life, and hence teaching and learning about sexuality.

A General Model of Work, Family, and Sexual Learning

Because the family and the work world are both embedded in the larger society, they influence and are influenced by social and economic conditions, institutions, and social norms. The most relevant of these norms are that men should work in the labor force and perform "breadwinner" roles for their families; that women's work is in the home, at least while children are young; and that men are not "supposed" to take primary responsibility for child care.

These norms are important factors leading to gender differences in the adult population in training, experience, job or career orientation, and time out for childbearing and rearing. They also contribute to different work opportunities for men and women, to occupational segregation in the labor force as a whole and at particular worksites, and to the general lack of adequate institutionalized child care.

Figure 3—1 presents a model of the impact of such a work world. It shows that parents must be viewed not only in their marital and parental roles, but also as workers. The work world contains both monetary and nonmonetary rewards; it demands time, money, and commitment from employees; it places them in particular work environments that have different demands and pay-offs and that lead to different kinds of learning. Finally, some work environments are gender-segregated, and some convey particular erotic messages. Thus parents' work roles make demands upon their resources, mental health, knowledge, values, and attitudes. Consequently, these factors

Figure 3–1. Connections Between the Work World and the Family and Sexual Teaching and Learning.

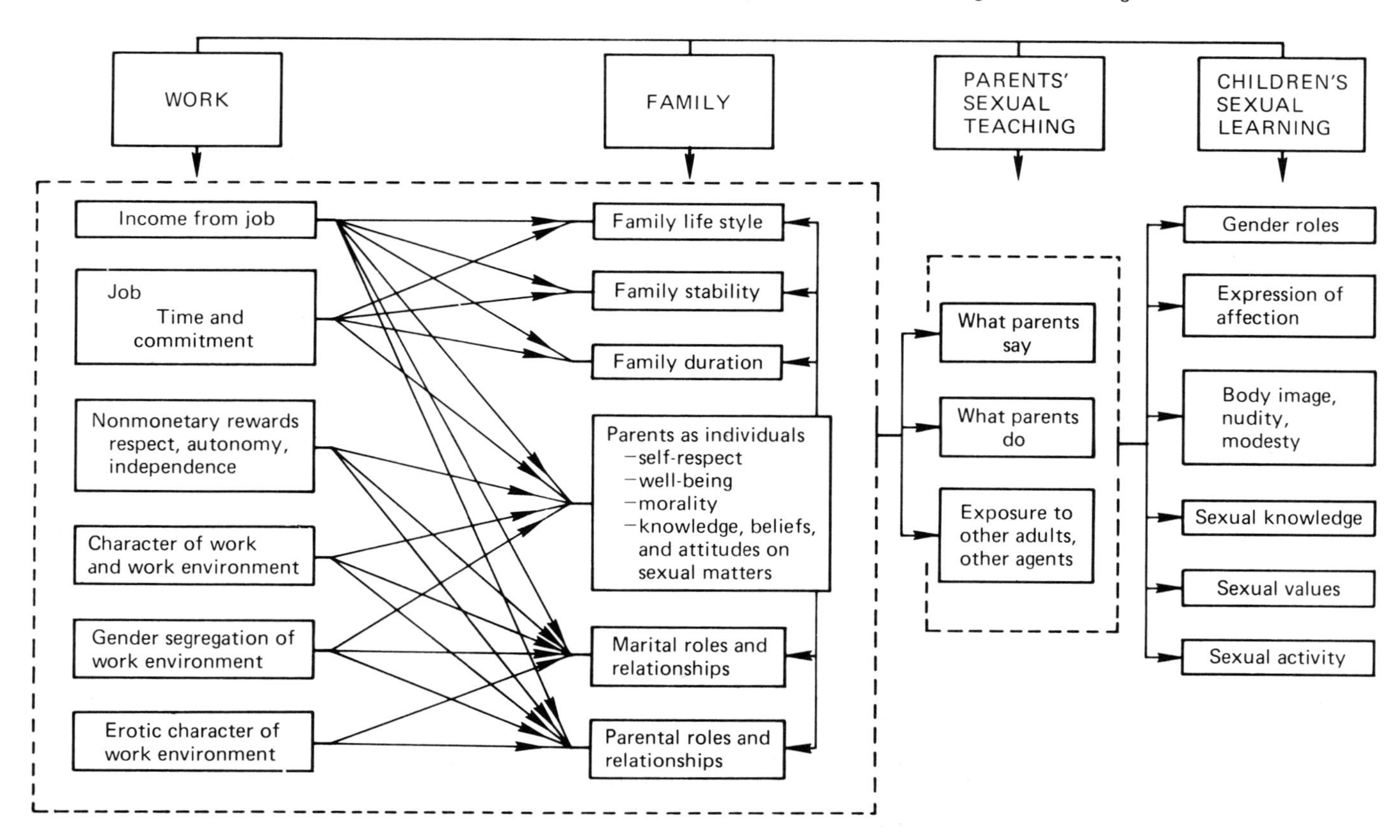

may also determine both how parents perform marital, parental, and nonfamily roles, and the lifestyle and stability of their families.

The work world may affect such family dimensions as: division of household labor; amount of time spent with the children—and what is done with them in this time; willingness to marry (and to stay married); number of children; social class; contact with relatives, friends, institutionalized child care, and school; harmony and conflict within the family; marital satisfaction for father and mother.

The right hand side of the diagram presents further possible consequences. Parental socialization of children depends on the parent's knowledge, beliefs, values, and self-esteem, and what is communicated by words and actions to children. Thus individual characteristics of the parents, their marital and family roles, and the systemic characteristics of the family affect the teaching given the child. Finally, this teaching comprises one source of children's learning about gender roles, expressions of affection, body image, nudity and modesty, sexual knowledge, sexual values, and the child's sexual activity.

Many of the arrows in Figure 3–1 should be double-headed; that is, family factors may affect work world factors, and children's sexual learning may affect the character of family life. This discussion will only examine correlations and relationships in which the direction of causality seems to lead toward children's sexual learning.

Investigations of the impact of the work world and the family on children's sexual learning must look at combinations of these factors, and must test (rather than assume) the causal connections. The complexity of such analysis is enormous, and it is scarcely surprising that no studies have traced the factors down the full chain, including variables from all four domains (work, family, teaching, learning). In few cases are there even speculations about relationships along the chain.

Individual studies reviewed for this essay each deal with only part of the causal chain—with the majority dealing with either the work world-family link or the family-children's sexual learning link. The former, however, has received very little serious attention from social scientists, (as other writers have noted), and those studies that do address the connections do not draw the implications for children.[1] Even the mobility literature rarely specifies intrafamilial variables; Peter Blau and Otis Duncan, for example, do not deal with the way in which a father's education and occupation are transmitted through the family to result in children's educational and occupational attainments.[2]

Literature dealing with the second link is generally theoretical and focuses more on sexual teaching than on sexual learning—often assuming that if we know about the former, we automatically know about the latter. Limited empirical research exists on what parents do and say, or on what children learn.[3]

This chapter reflects both the split and the lacunae in the literature, mixing reviews of existing studies, comments on the absence of data, analysis of possible linkages, and speculation about what might be found if good studies were conducted. An attempt has been made to clearly distinguish between these different approaches.

In Part II, the general characteristics of parental sexual teaching and children's sexual learning in the family are summarized. Parts III and IV contain analyses of the connections between work and family life. The theories and data in these sections will be used as a basis for generating hypotheses about the ways in which the work world may influence family life and children's sexual learning. Part V presents an agenda for further research.

Preliminary Clarification

Before turning to the first set of materials, three points must be noted. First, in common parlance and in much of the literature, the term "work" is used interchangeably with "labor force participation." Because they are not paid, persons (most often women) who spend fifty to sixty hours per week in household management and maintenance, and child care do not "work" and are not counted as part of the labor force. I have tried to use the terms "employed" or "labor force participant," reserving the term "worker" to include those who engage either in paid activities in the labor force or in unpaid work in the home. I believe this is not only more accurate, but it avoids such bizarre statements as "mothers who work do X" as if women who take full-time responsibility for their children and home do not work.

Second, the term "family" must give us pause. Most writings on the family's influence on child development really are examinations of the parents' influence, not of family influence. Furthermore, two parents are assumed to be present or at least desirable, and the one-parent family is ignored or is treated as problematic. The consideration of family influences on children's sexual development includes siblings, grandparents, aunts, uncles, and even friends who are treated as family members. Any or all of them may have child care responsibilities or may influence children by example.

In addition, I do not consider the one-parent family to be necessarily problematic, and I urge that we consider the many children who have more than two parents or more than two people performing parental roles. Rising divorce and remarriage rates mean that more and more children have in their lives other adults who have relationships with their parents. Mother and father may divorce, and both may remarry; the child becomes a member of two nuclear families, even though he or she may reside most of the time in one household. Both families are likely to affect the child's sexuality.

The third point relates to the work-childhood link. Even in our post-industrial society, some direct linkages exist between children and the work world. Children see doctors, teachers, mailmen, delivery men, storekeepers, salespeople, and truckdrivers at work. Sometimes they are just observed; sometimes there is interaction. The picture the child gets of the character of the work world from such contacts is probably very hazy; but whether it is accurate or inaccurate, it may be important in the child's perception of gender.

Other people are seen at work through mass media. On television children see fictionalized representations of workers, as well as newscasters and sports figures who are actually engaged in their work. The occupational distribution of television characters is very different from the real world distribution.[4] Newspaper accounts show some of the work of government officials and others. Magazine advertisements include pictures of people at workplaces.[5] However inaccurate mass media presentations may be, the work world they present may be influential in shaping children's sexuality, especially through gender-role learning.

Consideration of these three points is beyond the scope of this essay, but the omission is important to note and correct in future analyses.

II. SEXUAL TEACHING AND LEARNING IN THE FAMILY

We begin with an overview of sexual teaching in the family. In several domains parents act deliberately, serve as models, or offer explicit instruction.

Gender Role Teaching in the Family[6]

Although children in the United States (as well as in most modern industrial societies) are told little about sexual activity and the sexual roles that they may ultimately perform, they are told a great many things about who they are and what they are to do because they are boys or girls, and about what will be expected of them as men and women. The different treatment and teaching given to boys and girls of all ages can be seen in almost all families.

Sex-typing begins even before the child is born. Parents spend hours speculating about whether it will be a boy or a girl, and often refrain from purchases other than "neutral" items until knowledge of the infant's gender indicates whether the items "should be" blue or pink, and whether football players or ballerinas will be "appropriate" decorations. When the child is born, its gender is the first characteristic that the doctor mentions to the new parents, and that is communicated to relatives and friends. This announcement elicits a series of gifts that are frequently gender-specific.

During the first six months, mothers tend to look at and talk to girl infants more than to boy infants and to respond to girls more quickly when they cry. On the other hand, boys receive more touching, holding, rocking, and kissing than girls do in the first six months. By the time they are a year old, female infants are allowed and encouraged to spend significantly more time touching and staying close to their mothers. They are encouraged to move away at later ages, but never as much as boys are.

Robert A. Lewis suggests that the parents' interest in building autonomy or independence seems to play an important role in explaining these differences.[7] As a function of societal stereotypes, mothers believe that boys should be active, independent, and encouraged to explore and master their world. Thus sons are weaned from physical contact at an earlier age than daughters are. Other studies have also shown that parents put narrower limits on the acceptable behavior of infant daughters.[8] Daughters are treated as delicate, sweet, and dependent; when they behave accordingly, they elicit more parental approval.

Parents are not necessarily teaching gender roles deliberately. Having internalized society's general views of males and females, parents pass on these views almost automatically—unless some experience jars them into questioning their own gender views.

Sex-typing increases as the child moves from the infant to the toddler stage, at about age two. The process is helped by the child's increasing command of language and the parents' increasing use of it to mold behavior. Neutral rattles and rubber ducks are replaced by toys designed and labeled specifically for boys or girls. Most "boys' toys" are meant to be actively played with; girls' toys are either meant to be passively looked at or are directed toward house and child care. Girls are given dolls and dollhouses; boys get trucks, airplanes, and building sets and are told that they are "sissies" if they want to play with girls' toys. The division is clearly arbitrary, however, for it has been observed that many two-and-a-half-year-old boys would prefer the dolls.[9] Their parents encourage them to gravitate toward the "boys' toys," and children soon display different preferences. The shift is quite stable by the age of five or six.

At the same time, parents also mold their children's appearance and attire to fit society's definitions of masculinity and femininity. They comb and decorate a girl's hair, clothe her in dresses, give her jewelry, and let her play with mother's makeup and jewelry. They tell her not to get dirty, not to lose her pretty things, and not to do anything that will let her underpants show. She is taught to capitalize on good looks and coyness and to seek assurance from others. A boy is taught almost the opposite.

Children also begin to sex-type each other and themselves. They pick up and apply the labels that parents have used. If young children are asked how boys and girls differ, they agree that boys are noisy, interested in being outdoors, and dirty; girls are clean, quiet, gentle, and cry a lot.[10] If asked about the future, children see males as becoming protectors, adventurers, laborers, and decisionmakers; women will be cooking, sewing, and taking care of children.

By about age five, a child already has a firm gender identity and has learned a great deal from his or her parents about appropriate behavior. At this point, the child enters school, and adults and children from outside the family and neighborhood contribute to definitions of masculinity and femininity. In addition, the child is exposed to more books and to other mass media.

For some children, increased exposure to nonfamily influences will mean challenges to the rigid sex-typing that was learned in the family. Other children who have been exposed to minimal sex-typing in the home are likely to find more rigid definitions of appropriate gender behavior in the schools. A girl's independence and rough play, for instance, is not likely to be as encouraged in many schools as it may be in a nontraditional home.

Although the family continues to be important during this stage, family messages become part of a larger set. When outside messages are different, the family may find itself in competition with the mass media, schools, and peers. On some counts parents will feel that they have "lost." For example, a boy who was raised to be warm, affectionate, and gentle, may return from school acting very "macho" and announcing that he hates girls and girls' activities.

The family is also of importance at this point because it controls the child's access to other agents. If they can afford a private school, parents may remove a child from a public school that is teaching the "wrong" morality, permitting or encouraging "improper" gender behavior, or ignoring "inappropriate" sexual conduct. Parents not only choose the religious institution that the child will attend, but also govern the frequency of attendance. Finally, *some* degree of control is exercised by parents over the peers and media experiences of their children.

Expressions of Affection, Modesty and Nudity, Body Learning

Children also learn in their family environments about physical affection, modesty, and nudity. Children's senses of their bodies and their capacities for physical pleasure also begin in the home at an early age as they are held, suckled, caressed, and fondled—or as they receive little of these physical contacts. In later childhood, they continue to receive both explicit and implicit teaching about the appropriateness of touching, hugging, and kissing.

Families also vary in the normative and behavioral environments that they create for nudity, and in their responses to violations of the family norms. Children thus learn about whether, under what conditions, and to whom it is appropriate to show one's body or certain parts of it.

In some cases, regular and consistent patterns are developed. Family members either do not touch each other very much and are not nude in front of each other, or they have much physical contact and are at ease about nudity. In other households variations exist by gender (girls can see the mother nude; boys can see the father) or by the age of the child. By the time children are near age eleven, cross-sex nudity between parents and children may not be permitted (if it ever was), and physical affection between males may have decreased. For example, a young girl may receive much affection and caressing and may freely encounter her parents nude, but as she draws closer to puberty, her father may feel increasingly uncomfortable and may fondle her less and be sure to dress when she is around. (She, of course, may be the one to initiate the change.) Likewise, when he is a toddler, a boy may be encouraged to hug and kiss everyone at bedtime; suddenly one day as a young boy he may be told to shake hands with his father and brothers because "boys don't kiss boys."

General Knowledge

Genital Information. Just as young children ask their parents seemingly endless questions about various parts of their world, they ask questions about genitals. First, names are wanted—then further information.

Parents respond to such questions in a variety of ways. Some defer and dodge; others answer in baby-talk; still others give children the correct technical terms.

Parents may also use pictures or books to show children the differences between men's and women's genitalia. Because few such materials are available, however, this method is probably still infrequently used.

The phrasing of answers about male and female genital differences is also important. Often the answer concerns why boys have penises and girls don't. Few parents seem to explain more fully that boys have penises (and girls don't) and that girls have vaginas (and boys don't). Even fewer include mention of the clitoris. Hence, many supposed discussions of genital differences are really discussions of male genitalia.

Lovemaking and Reproduction. Discussion between parents and young children (under age eleven) about sexual activity is usually focused on reproduction. It starts either with a child's question about where babies come from, or less frequently with an adult's decision that the child should learn the facts of life.[11]

Again, parents deal variously with the topic. Some defer questions to a later time; some offer the fiction of the stork; some use a parable about flowers and bees, and hope that the child will see the connection to adults; still others answer factually. The answer will depend partly on the child's age and greatly on the parents' attitude toward what is appropriate.

Even where a factual account is offered, the starting point is significant. Many parents—like many books on reproduction designed for children—begin with the sperm and the egg, and neglect to explain how such things came to be in the same place. Rarely is there a discussion of lovemaking as the prelude to conception, and even more rarely is there a discussion about lovemaking other than in the context of reproduction. Much learning about sexual activity must depend, therefore, on other sources—especially peers and the mass media ("dirty books," magazines, and films)—that may often contain erroneous information and convey different attitudes from the ones parents would like their children to develop.

Sexual Values and Sexual Activity

Between the ages of three and eleven, children learn not simply some version of what sexual matters are and how they happen, but they begin to develop a set of attitudes toward sexual matters. More fundamentally, they develop a preliminary set of sexual values—conceptions of the desirable. These values are preliminary in the sense that they are subject to later modification in the light of additional information and exposure; yet values that are deeply rooted may serve to deter later exposure to contrary sources. For example, the child who learns that homosexuality is terrible may later avoid any contacts that would challenge this view.

Sexual values are formed around what John Gagnon refers to as "script" elements: who one should have sex with (age, gender, and relationship to you), where, when, why, and under what conditions.[12] The script learned by many children is a narrow one: "sex is for mommies and daddies—with each other, in bed at home, once in a while, to have kids." Other children may have broader versions: "mommies and daddies can do it more than once in a while;" "it's OK for adults who are in love;" "it's an activity for fun, and because it feels good."

To some degree children learn these values through parents' direct statements. Many parents, however, have difficulty in talking about sex at all. Others who can talk about it seem to fear that giving children too broad a "script" will lead to their experimenting "too early," or they think that "those things are OK, but not for *my* child." The double standard for daughters and sons also persists.

Thus, parents often communicate to their children a more rigid and narrow set of standards than they themselves hold. Of course, in the rapidly changing world, many parents may find themselves confused about or changing their own sexual values. This confusion, however, seems rarely to be acknowledged to children, though it may be communicated to them through unclear or conflicting messages.

Children are also taught about sexual values by what parents say in the course of conversations with one another, with other adults, or with siblings, as well as by direct and indirect messages from other family members. Children may make judgments about what is sexually good and bad or right and wrong from listening to what is praised, damned, ignored, laughed at, laughed about, or from noticing what is the butt of a joke, the source of teasing, or the source of joy or rage.

Labels and language may also not be understood. Terms such as "queer," "fag," "a good lay," "fast," or "give head" may be heard from parents or siblings without any understanding. Later, when the meanings are learned, the emotional tone may be remembered and may influence judgment.

Children also learn something about sexual matters from parental avoidance of discussion. Thus children may infer that there is something wrong with talking about or even thinking about them. Similarly, parents who do tell their children about sexual matters before most other children have been told—or parents who have given their children more "liberal" messages about sex—may conclude the talk with an urging not to discuss these things with friends. Whatever the explicit message about sex, the covert character of sexual matters may be reaffirmed.

Finally, much of the child's learning about sex comes not from what is said by parents but from the emotional tone, anxiety, discomfort, fear, or moral concern that often accompanies parental response to questions. Even greater doses of these emotional responses may accompany parents' discoveries of their children's "sexual" activity—what seems to be masturbation or instances of children's sexual play.[13] Children may not understand the parents' response, but they are aware of its potency.

The Influence of Other Family Members

The foregoing description of sexual teaching and learning alerts us to the necessity of examining such factors in the family as the content of sexual teaching, the amount of it, the ease or difficulty with which communication on sexuality takes place, who initiates discussion, and the credibility of the source. We have also seen that we must ask about both what is *said* and what is *done* with respect to gender roles, intimacy, modesty, nudity, lovemaking and reproduction.

The foregoing section, however, has focused primarily on the influence of parents. A full analysis of the influence of the family requires that we also ask the above questions about siblings and other family members. The presence of siblings means that there are other models and other sources of ideas, data, attitudes, and behavior. Siblings may be more likely than parents to be nude and "immodest" and to talk of their own sexual behavior. They also are representatives of "youth culture" within the house. We must ask, therefore, about the number and gender of siblings and the birth order. Grandparents or aunts and uncles who spend much time with children may also have a marked impact. Since 60 percent of the children of black working women and 40 percent of the children of white working women are cared for by relatives, the influence of these family members cannot be ignored.[14]

Finally, as suggested in the Introduction, the current high divorce and remarriage rates mean that many children are members of more than one nuclear family. Instead of asking only about the teaching by the one or two parents with whom the child lives, we may need to ask about the teaching by three or four parents.

The presence of these additional sources of family influence on children's sexuality creates considerable analytic confusion. We do not yet know what are the effects of conflicts in the proscriptions, prescriptions, and examples of more than one family member or of more than one family. We must not hastily assume that greater time spent means more impact. It may be that a scarcely seen uncle teaches more about sexuality than parents who won't talk about it, and that the gender modeling of a parent with whom the child does not live is more instrumental in the child's gender-role development than the parent with whom he or she spends the most time.

III. WORK INPUTS AND OUTPUTS AND THE FAMILY

Having sketched the ways in which parents teach their children about sexual matters, we now come to the influence of the work world. The Introduction has proposed that what parents say and do in their sexual teaching is partly a function of their relationships to the work world. Their job experiences are critical in terms of the resources of time, energy, and income that they have available for their families. Thus, we look first at work inputs, and outputs and family life.

Inputs by the Worker

Time and Timing. Jobs differ as to the inputs required from those who occupy them. Some of these inputs take the form of prior training—whether in a formal academic program (such as medical school or business school), in a trade school, in an apprenticeship program, or in prior work of the same sort at a lower level of skill and pay. An initial dimension of the work time input, therefore, is the time spent qualifying for the job in the first place.

A second dimension of work time input is the number of hours spent at work per day, per week, and per year. Jobs that involve the same number of hours per week may not require the same number of hours per day. For example, a truck driver may spend forty hours a week on the job, but the hours may be bunched into three days of intensive work. An accountant, on the other hand, may spend the same number of hours, but the hours will be spread over a five-day week. Similarly, people in seasonal occupations may spend vast amounts of time at work during the "season," and then have slack periods of little work. We must ask, therefore, about the amount of time spent at work; the regularity of work hours; when during the day, week, or year the time is spent; and how much control the worker has over the time spent.

For many blue-collar and white-collar workers, the time spent at work is fairly regular. Today, approximately forty hours per week (spread over five days) is typical. However, many business executives regularly work at least sixty hours per week,[15] as do many college professors and other professionals. Much of this time is spent on work taken home, rather than done in the office.

Again, for many people the work day occurs roughly between 9:00 A.M. and 5:00 P.M. Blue-collar workers and salespeople—both of whom may be assigned to early morning or evening shifts—are more likely to work on different schedules and to have rotating weeks. Other jobs (such as teaching school) that are not generally considered seasonal have periods of nonwork and therefore subject the workers to a less-than-regular schedule.

Workers who do not work at full-time jobs may work part-time; or workers may work overtime or have two jobs. At the other end of the spectrum, less than 10 percent of married working women hold full-time jobs for the whole year. Three-quarters are in part-time employment, half of these for only part of the year. The remainder of working married women take on full-time employment, but for only a fraction of the year.

The final work time input is flexibility. To what extent can the worker choose when and how much he or she works? In general, the higher a worker is on the class ladder, the greater the control he or she has over time spent. Similarly, men have more control over their time than do women, many of whom have no alternatives to taking part-time employment, with its accompanying lack of benefits and seniority.

Physical Energy. Related to work time is the degree of physical fatigue resulting from that time. While most jobs in our modern, technological, industrialized society do not depend on physical strength,[16] some are more demanding than others. Robert Blauner suggests the auto assembly line is one of the few factory jobs where heavy physical labor is still required.[17] Fatigue may also stem from temperature variations or lack of ventilation, from standing most of the day (as in much saleswork), or from climbing up and down stairs.[18]

Absorptiveness. A further dimension of the energy input to work is what Rosebeth Kanter has called "absorptiveness," that is, the degree to which one's work commitment pervades the rest of his or her life and the degree to which it involves other family members.[19] In the first instance, a job may carry with it implicit or explicit demands that the occupant engage in "unofficial" activities—such as community responsibilities, club memberships, or socializing with colleagues. Many researchers have described the "all encompassing" character of many high positions in business and politics.

In the latter instance, absorptiveness refers to demands on the time of other family members. This has been described in considerable detail by Hanna Papanek, who writes of the "two-person single career."[20] Although one person (the husband usually, but occasionally an "exceptional woman") is employed by an institution, formal and informal demands are made upon both members of a couple. Examples are corporate executives, university and government officials, ministers, and officers in the armed forces and diplomatic service. The wives of these workers are expected to reinforce their husbands' commitments to the institution, and to fulfill such functions as hostessing, attending company affairs, engaging in volunteer work, and joining in special clubs and activities for wives.

This kind of absorptiveness is not in and of itself bad. Many executive wives, for example, would describe their involvement in the social activities of their husbands' careers as a benefit. Wives may enjoy the recreational, social, and political dimensions of such informal responsibilities, and may prefer engaging in these tasks to taking poorly paid employment. Wives and children may gain travel and recreational opportunities (such as country club membership) from the husband's or father's work.

A point to be noted is that in many cases the tasks of the wife are obligatory (or close to it) if the husband is to survive and thrive within the institution. She may have little choice, and may refrain from paid employment altogether. Or she may restrict or divert her own professional ambitions to avoid conflict with her husband's work requirements. Furthermore, wives are paid for these tasks only indirectly and—despite heavy investments of time and energy—wives are considered people who "do not work."

Inputs to Work and Family Life. The time, timing, physical energy and absorptiveness of work also have implications for family life.

The contemporary family is usually heavily organized by the rhythms of the work world. Ordinarily workers leave in the morning, return at night, and have the weekends off. Changes of these patterns, as when there is night work, extensive travel, or fixed tours of duty away from home, disrupt the usual family routine and require some adaptive response from the family.

Even when people have more time not spent at work, however, or when they have more control over their time, they do not necessarily spend more of it in family ways. Time spent at work is time that cannot be spent with the family or at leisure activities. It is not at all clear how much time, if freed from work, would be devoted to the former rather than to the latter.

In looking at the impact of the work time and timing on family life, we must once again recall that we cannot look only at the situation of men and the situation of women, but must look at the situation of couples.

For men, work responsibilities tend to limit the time they have available for their families. For women, on the other hand, the relationship is more symmetrical: work responsibilities limit time for family life, but family responsibilities limit time for work. Since her husband's job is usually conceived of as the "first" job, her schedule often has to fit his. If he leaves at 6:00 A.M. for work, she is likely to seek a job that doesn't require her to leave home until the children are dispatched to school. If he is not at home by midafternoon and no child care is available, she may feel the need to get home from work early. The husband's job times, therefore, may create the time frame within which the wife's job can be set. Such a restriction places severe limits on the job options available, and often creates the need for part-time and part-year work.

Exceptions to this general pattern, of course, do exist. Single parents, who may need to work long hours in order to make ends meet, will have less choice about how to manage both work and family. In some communities, because child care facilities are well developed (an uncharacteristic situation[21]) mothers may be freer to pursue work goals in the labor force. In some families, child care responsibilities may be equitably distributed between husband and wife, or allocated to a third party.

The absorptiveness of work is also different for mothers than for fathers. The cultural bias is obvious as soon as we look at the degree of absorptiveness in work that is considered problematic. It is sometimes considered a problem if the husband-father's work takes him away from the family for much more than the usual forty to fifty hours a week. Thus we hear laments about the business executive or physician who, though earning a handsome income, is a "workaholic" or "neglects" his family because of heavy work responsibilities.

For the wife-mother, the problem of absorptiveness arises as soon as the question of labor force work is considered. While her children are very young, a mother is urged to avoid paid work on the assumption that her time away from the children will be bad for them. Her husband's time away is not considered a similar problem, because it is the mother, not simply a parent, who is considered necessary for the children.

The assumption here does not have to do with the male's higher earning power, but with the "naturalness" of the mother's child care responsibilities and the "naturalness" of the father's labor force work. Men who have elected to stay home and be "househusbands" report negative responses from friends and neighbors.[22] Even if neither parent needs to work for income, social opprobrium is cast upon the man who stays home and the woman who goes out to a labor force job.

The problem of absorptiveness also tends to be greater for men than for women because currently absorptive jobs such as corporate executive or government official have largely been jobs for men. As more women come to occupy these high status, time- and energy-consuming positions in the labor force over the next few decades, the same problems that men have in these occupations will very likely arise for women. Unless their husbands become "househusbands," their problems will become those of two-career families. As available data is limited, the impact is at present largely a topic for speculation.

Outputs to the Worker and the Effects on Family Life

The Effects on Family Life. The inputs and outputs to and from work affect family life in a variety of ways—including the family's stability and rhythms, and the amount of time spent in household tasks and childrearing. Each of these variables may be expected to have consequences for what parents teach children about sexuality. In general, these work factors probably have more impact on what parents do than on what they say. In the following pages, we shall examine the ways in which life-style, stability, and marital and parental roles are shaped by the work world. A major set of outcomes involves what parents demonstrate to their young about the nature and significance of gender differences.

Men's Work and Income and Their Relation to Family Life. An obvious connection between work and family life in modern societies comes through the job income. Historically, men have been considered responsible, in both the normative and the legal sense, for providing economically for their families. Being the "sole provider" is still felt to be a mark of personal worth and success by many men.

The substantial variation in the resources provided by jobs, however, means that there are differences in the extent to which men can provide for their families. Many men who work—even those who work full-time, year-round—do not earn enough to provide adequately for their families' needs. Depending on the standard used, between 10 percent and 30 percent of families are estimated to live below the poverty line.

Connected to occupational categories are also different patterns of income increments. Blue-collar work (particularly factory work) commonly reaches salary maximums rather quickly. Hence, after his late twenties the blue-collar worker often cannot anticipate doing much better than "keeping even" with inflation, though the economic needs of his family are likely to grow substantially as his children grow older.

Among middle managers, according to Emmanuel Kay, annual rates of salary increase tend to be highest early in a career and then to level off at about age thirty-five or forty.[23] In many other white-collar occupations men reach income peaks in their early and middle forties. They often will obtain salary increases then, at a time when their children make additional demands on money for social and educational endeavors.

In some cases, the solution is for the man to work more than one job. Lee Rainwater reports that young working-class men often cope with the problem of inadequate income in the early years of family-building by seeking extra jobs and overtime.[24] Schedules that require ten to fourteen hours a day at work will bring their incomes closer to the median income and permit a less marginal existence, but will also be fulfilled at great costs in time away from home.

Male income may affect the initial family formation. In this sense, several researchers have suggested that the timing of marriage is related to work and income. Anticipated higher income may lead to deferred marriage—as among men going to college or graduate school.[25]

Others have suggested that whether families are formed at all, and what type is formed, may be related to work factors. Men with a high investment in their careers or future careers are more careful in their selection of sexual partners and more scrupulous about using contraception when they engage in premarital sexual relations.[26] On the other hand, low income potential may preclude some young men from marrying even when a child has been conceived. Similar reasons may prevent many welfare mothers from remarrying.[27]

The stability and duration of family life is also related to male income. Low-income jobs mean inadequate financial resources and all the problems of poverty. Such families must plan expenditures exceedingly carefully. Apart from the struggle for daily needs, there may be no cushion for errors of judgment or unexpected crises. In many cases the marriages cannot withstand such pressures, and marital disruption—either desertion or divorce—is the result. Indeed, the divorce rate is considerably higher for couples with lower incomes.[28]

Even where the marriage remains intact, the low-income family is likely to experience considerable tension and problems. Karen Renne found that income was more closely related to marital dissatisfaction than education or occupation. She has argued that the domestic problems of black families can largely be attributed to economic deprivation.[29]

Since the poor family is much like most families in the society in that it is not composed of remarkable individuals, members are often unable to cope with the psychological costs of being poor. Rainwater reports that low income leads working-class men and women to pay considerable costs in ever present anxiety.[30]

Women's Work and Income. American social values have endorsed the notion that the proper role for women is to be found in the home. Since colonial days, however, American women have labored on farms and as shopkeepers, doctors, midwives, nurses, and teachers. When industrialization arrived in the nineteenth century, large numbers of women found employment in the mills and in canning and tobacco factories.[31] Women's Bureau studies (the Women's Bureau is part of the U.S. Department of Labor) show that in the 1920s and 1930s approximately 90 percent of employed women went to work because of economic need and used their income to support themselves and their families.

During World War II female employment became a national necessity, and in four years more than six million women entered the labor force—including many who were married and over thirty-five. The major shift since that time has been in the employment of married women. While the percentage of single women who work in the labor force has remained constant at about 50 percent for the past thirty years, the proportion of married women has grown almost 300 percent from 15 percent in 1940 to 40 percent in 1970.[32]

In some instances today, working women are heads of households and seek work out of simple economic necessity. In other cases, married women have entered the labor force because their husbands' incomes are inadequate to support the family. A wife's employment may also be a means to achieve and maintain a middle-class standard of living. In yet other cases employment is largely motivated by the noneconomic rewards provided by labor force participation (see Section IV below).

Household Work: Unpaid Labor. Participation in the labor force is one dimension of the work commitment of the family. The household also involves large amounts of work, including household duties and child care. But this work, unlike labor force participation, is unpaid (at least when performed by a family member). Though it clearly has an economic value, economists have not agreed upon a manner of calculating the value of time spent in housework.[33]

The unpaid character of household work has particular significance in family life because household responsibilities do not tend to be distributed evenly between husbands and wives or between parents and children, but rather tend to be almost the sole responsibility of women. Such an arrangement seems reasonable to most men and women in our society, for they have been socialized to believe it is "natural."

Where women find the arrangement burdensome, they are impeded in changing it not only by the lack of adequate institutional child care facilities but by labor force discrimination. The fact that in most families the husband can earn more money through labor force work makes it an economically unsound decision for him to quit his job and care for the house and child. (This should not mask the fact that values other than the economic ones can easily be weighed against the economic loss of such an arrangement.)

Thus, where there are two parents and only one has a paid job, it is most likely to be the father who earns money and spends time at a job. The mother who "does not work" in official parlance, however, is likely to have almost sole responsibility for household tasks.

Income Differentials within the Family. In a large number of American families the husband is still the sole breadwinner. Where wives do work, they generally have lower incomes than their husbands—contributing between 25 and 30 percent of the family income.[34]

Several reasons can be found for the disparity. First, there is considerable sex discrimination in the labor market. Even where researchers have standardized for education, industry, and occupation, they have found a substantial male-female disparity in income that can be accounted for only by discrimination.[35]

Second, women tend to marry men who are slightly older than themselves and who have somewhat more education. Thus, even if there were no sex discrimination, wives would earn less than their husbands because of their lesser amount of training and experience. Third, many women interrupt their work lives when their children are young, and thereby increase their relative disadvantage. Finally, many married women with children work at jobs that are less than full-time positions—losing both income and fringe benefits.

When Both Parents Work. When there are two parents and both hold labor force jobs, the situation changes somewhat for women—who spend significantly smaller amounts of time in housework, but still spend many hours at it. For example, Kathryn Walker[36] found that employed women averaged four to eight hours daily on household tasks, while unemployed women averaged five to twelve hours daily in housework. In seeking an explanation for such differences, Joann Vanek[37] tested the hypothesis that women who were not employed outside the house had larger families and more young children. She reports that although controlling for these factors reduces the time differences between the two groups of women, the major amount of difference remains.

When both parents work, however, the allocation of household responsibilities does not change much for husbands. Contrary to popular belief, even when their wives work, American husbands do not share very much in the responsibilities of household work. Married women in a 1966 study spent an average of forty hours a week keeping house; married men averaged four hours a week.[38] A more recent study found that men averaged eleven hours per week in household tasks.[39] Differences in men's time in household tasks were related to differences in their hours of employment and were not correlated with employment of wife, number of children, or age of the youngest child.[40]

A few case studies (mostly of dual-career families) report husbands taking on tasks ("to help their wives"). Other studies of such dual-career families report that the household tasks of two-thirds of the couples were either performed by the wife with the assistance of hired help or remained the primary responsibility of the wife.[41] Virtually never is this as "spontaneous and natural" as some observers, including some social scientists, depict.[42] Some studies have found that children take on some of the tasks; others have found children's contributions negligible.

The mother who works, therefore, tends to have two jobs. The tradeoff for her market work is not household work, but a reduction in her leisure time.

In summary, the child who grows up in a home with two working parents is likely to see that the mother earns less money than the father. The mother still does most of the household work, and she has very limited free time. Very few children, therefore, grow up in homes where they see gender equality between their parents.

Since the child's learning about gender and about life-styles is likely to be influenced by the way parents model gender, we must ask what gender-related differences exist in the marital and parental roles of dual-career parents. Note that this discussion is often headed "the effects of mothers' work." That is actually a misnomer. The effects are effects of both parents working. The heading arises because of the tacit assumption that he *will* work and that she *may*.

Wives who work outside the home may have greater authority in family decisionmaking—especially on economic matters. In working-class homes in which a husband continues to believe in the "rightness" of his authority, the wife's authority may be a source of conflict. Lillian Rubin reports that in well over one-third of the families she studied husbands claim their working wives are getting too independent. Comments made to her by a thirty-three-year-old repairman are reported as typical:

> She just doesn't know how to be a *real wife, you know, feminine and really womanly.* She doesn't know how to give respect because she's too independent. She feels that she's a working woman and she puts in almost as many hours as I do and brings home a pay check, so there's no one person above the other. She doesn't want there to be a king in this household . . .
>
> I'd like to feel like I wear the pants in the family. Once my decision is made, it should be made, and that's it. She should just carry it out. But it doesn't work that way around here. Because she's working and making money, she thinks she can argue back whenever she feels like it.[43]

Other researchers have focused on the effects on other family members. Studying dual-career families, Holstrom found conflicts stemming from dilemmas of role overload, identity (masculine and feminine stereotypes), and role cycling at critical stages (as when a child was born or when one of the partners received a promotion).[44]

Susan Orden and Norman Bradburn found that if the woman's work was by choice rather than necessity, it had a positive effect on both the wives' and the husbands' assessments of their marriages.[45] It must be noted, however, that it is not possible with the current data to tell whether it is the effects of the wife's employment or self-selection (the particular personalities and attitudes of those who elect to go to work) that is more important in explaining the positive assessments. It may be that when both marriage partners work or have careers, they are from the outset more committed to relative freedom and independence. Such attitudes may lead to the wife's employment and—because the husband may feel less economic burden—to lower tension levels and higher satisfaction for the couple.

For similar reasons, it is very difficult to ascertain the effects of maternal employment. Working mothers differ from nonworking mothers in a number of ways. They tend to have fewer children, older children, higher education, husbands who are more supportive of a work role for wives, husbands who are somewhat more active in household and child care, residence in communities with more job opportunities for women, and friends with attitudes favorable to women's employment outside the home. Any of these variables alone might lead to differences between their children. Most studies, however, fail to control for such variables.[46] In addition, studies often exclude black families and single-parent families.

While some people continue to sound the alarm about the detrimental effects of the mother not caring for her children full-time, other people point to reasons to expect positive effects. They note that the working mother presents a better model of female competence and achievement, that the father is likely to take on a more active role in child care, and that there is likely to be less economic stress. Several studies lend some support to this line of reasoning, although the exact chain of causality is unclear. Children of working mothers have not been found to get less attention or to be emotionally deprived. The quality of the interaction between mother and child, rather than the quantity, is what seems to be critical. A key factor seems to be the mother's emotional state concerning her work; the children who suffer most are those whose mothers want to work in the labor force but feel considerable strain, tension, and guilt because they do not. Thus, Lois Hoffman reports that the working mother who obtains personal satisfaction from employment, who does not have excessive guilt, and who has adequate household arrangements is likely to perform as well as the nonworking mother or better.[47]

Finally, when both parents work, effects on marital duration may be noted. If economic instability leads to poor marital relations and higher divorce rates, and if wives' labor force participation shores up family economic situations, we would expect that rising work rates for wives would be accompanied by greater marital stability. As a general rule, however, Kristin Moore and Isabel Sawhill suggest the contrary. Increased labor force participation by women has been accompanied by rising divorce rates. The trend is due partly to tensions in the family from women's work, but mostly because unsatisfactory marriages that were held together by women's economic dependence no longer need endure. Thus, these authors predict that divorce rates are likely to continue to climb as more wives enter the labor force, but that those marriages that do last will be stronger.[48]

IV. WORK, PSYCHOLOGICAL WELL-BEING, ON-THE-JOB LEARNING, AND FAMILY LIFE

A second way in which the work world may affect family life and children's sexual learning is the quality of the experience that the parent-worker has at his or her job.

This quality has two dimensions. First, jobs characteristics vary: some jobs are repetitive and supervised while others have considerable variety and allow for independent decisionmaking. Objective job conditions are mediated through people's subjective experiences. What one finds boring, another may find challenging. Subjective responses to work—boredom, alienation, excitement, and involvement—contribute to the parent-worker's sense of psychological well-being. This sense in turn affects how that person functions as an individual, as a spouse, and as a parent.

Second, continuing socialization takes place at work. In work contexts, people may learn new approaches to morality in general (honesty, dependability) and may have experiences that affect their thinking, feeling, and actions regarding gender roles and sexual activity.

Work, Respect, and Self-Respect

The Psychological Importance of Employment. Because American cultural values stress the importance of the male breadwinner role, work constitutes a major source of identity for most men. Rainwater considers work to be men's primary "validating activity" in this society, and by pointing to the critical importance of simply having a job, he argues that psychological rewards are provided even by menial work.[49] Beyond this base of self-esteem, jobs "pay" individuals differently in terms of respect and prestige. To a large degree the prestige and respect generated by a job accord with its educational requirements and pay level; to some degree they are independent. Some people who work regularly complain that they are nonetheless looked down upon. Others suffer constant humiliation from their job roles, and are unable to invest the job themselves with more respect than others do.[50]

The importance of work to men's psychological well-being has been illustrated through a number of studies of unemployed men.[51] Unemployment is usually accompanied by serious problems of self-esteem and identity, strained family relations, and damaged physical health.

While labor force work has provided this critical element of identity and adult status for men, home responsibilities (in the traditional model) have provided validating activities for women. In a recent study, however, Daniel Yankelovitch found that some of the same identity correlates of labor force work are becoming important for women, as paid jobs become "badges of membership in the larger society" and "an almost indispensable symbol of self-worth."[52] Similarly, Rubin's female respondents refer to compensations (for working outside the home) that go beyond the material ones: "a sense of being a useful and valued member of society . . . of being competent . . . of being important . . . and of gaining a small measure of independence from their husbands."[53]

Occupational Segregation and Differential Respect for Men's and Women's Work. As women have suffered in income relative to men, so too have they suffered in terms of access to prestigious jobs. One of the characteristic differences between the work worlds of men and of women is the degree to which the latter are concentrated in a small number of occupations. In 1970, half of all women workers were employed in seventeen occupations; half of all men were in sixty-three occupations.[54] In recent decades the overall growth in female labor force participation has not been constant across occupational groups. There has been a drop in some fields, a disproportionately small increase in professional and technical workers, managers, officials, and proprietors, and a disproportionately large increase in clerical workers.[55]

Furthermore, the occupations considered "women's work" tend to be given lower esteem than the others. Extreme examples of this are found in a study by Mary Witt and Patricia Naherny of the way traditional women's occupations are rated in the *Dictionary of Occupational Titles* (DOT).[56] This volume describes and rates the level of complexity of over 30,000 job titles, and is used as a basis for determining compensation and perquisites.

In the DOT jobs are given three-digit ratings. Each digit represents the complexity of the job, on a scale of zero to eight, for one of three categories—data, people, or things. A middle digit of zero, for example, indicates the highest level of complexity in dealing with people is required. Thus the rating of 101 given a surgeon indicates high skills are required in all three categories. Women's jobs in general receive extremely low ratings. Some examples are offered in Louis Howe's review of the study:

> Foster Mother (dom. ser.) . . . 878. "Rears children in own home as members of family. Oversees activities, regulating diet, recreation, rest periods, and sleeping time. Instructs children in good personal and health habits. Bathes, dresses and undresses young children. Washes and irons clothing. Accompanies children on outings and walks. Takes disciplinary action when children misbehave. . . . May work under supervision of welfare agency. May prepare periodic reports concerning progress and behavior of children for welfare agency."
>
> Rating almost but not quite as low as foster mother was the following occupational title, which received a slightly higher 874:
>
> Horse Pusher (agric.) . . . 874. "Feeds, waters, and otherwise tends horses en route by train."
>
> Child Care Attendant, 878. " . . . House parent, special school counselor, cares for group of children housed in . . . government institutions."
>
> rates the same as
>
> Parking Lot Attendant, 878. " . . . parks automobiles for customers in parking lot . . . "
>
> Nursery School Teacher, 878. "Organizes and leads activities of children, maintains discipline . . . "
>
> much less complex than
>
> Marine Mammal Handler, 328. "Signals or cues trained marine mammals . . . "
>
> Nurse, Practical, 878. " . . . cares for patients and children in private homes, hospitals . . . "
>
> only slightly less complex than
>
> Offal Man, Poultry, 877. "Shovels ice into chicken offal container."
>
> Home Health Aid, 878. "Cares for elderly convalescent or handicapped"
>
> about as skilled as
>
> Mud-Mixer-Helper, 887.
>
> Nurse, Midwife, 378
>
> nearly as skilled as
>
> Hotel Clerk, 368.
>
> Homemaker (cross references Maid, General), 878
>
> nearly as skilled as
>
> Dog Pound Attendant, 874.[57]

These findings attest in a most appalling way to the differences in respect accorded men's and women's work. We shall return to this difference when we look at how work may affect the messages that parents give their children about gender differences.

Work and Psychological Well-Being

Job Satisfaction and Psychological Well-being. If employment is the critical determinant of psychological well-being in the work-world, it is not the only one. Jobs and work environments vary in many ways other than their monetary and esteem rewards—including opportunities for advancement, job security, physical conditions of work, work load, responsibility, pressures, clarity versus ambiguity of requirements, autonomy, supervision of others, monotony versus variety, and various dimensions of social relations with others.

Such elements are important because they often affect the way that workers feel about the job and also about themselves. Ethnographic accounts, such as Studs Terkel's *Working*, are filled with descriptions of differences between jobs and the resulting emotional responses of those who perform them.[58]

It is impossible within the scope of this paper to summarize all that is known about the links between job characteristics, emotional responses to them, and psychological well-being. Table 3–1 presents an outline of some of the major findings, and the interested reader is referred to the extensive summaries of studies in James O'Toole's *Work and the Quality of Life*.[59] The present discussion is limited to conclusions that might be drawn from existing studies.

First, we do not really have a good understanding of what creates work satisfaction or dissatisfaction. People, of course, differ in what they desire from jobs. Sometimes rather than studying differences in what leads to satisfaction, researchers have made assumptions about preferences or tolerances based on gender. In the textile industry, for instance, it has been said that women are satisfied with low-skilled, highly repetitive jobs "because women are more content than men with work that is of little challenge."[60]

Second, even when a worker admits to a dissatisfaction with several things in a job, he or she may be "satisfied" with the work overall because of one strong, overriding, positive element.[61]

There is also a continuing debate in the literature about how much work dissatisfaction actually exists and whether it is growing.[62] Some researchers argue that job dissatisfaction is a widespread problem;[63] while others echo the July 1978 report by the National Industrial Conference Board, which concluded that "the vast majority

Table 3–1. Summary of Studies on the Quality of Work, Life and Family Life.

Study	*Work Characteristics*	*Responses*	*Family Life*
Kornhauser, 1962	High skill and interest level required	"Good" mental health	
McLean and Taylor, 1958; Eaton, 1967; Steiner, 1972	Internal mobility of executives within organizations	Loneliness; inadequacy feelings; negative consequences of stressful competition and constant change	
Kahn and French, 1962; Kasl, 1974; House, 1974	Low paid work, little authority	Stress, alcoholism	
Margolis and Kroes, 1974	Role ambiguity, role conflict, role overload, responsibility for others, poor relations with others	Coronary heart disease	
Palmore and Stone, 1973	Continued employment	Work satisfaction, longevity	
Margolis and Kroes, 1974	Role overload	Low job satisfaction, coronary disease, high serum cholesterol levels, high heart rate	
McKinley, 1964	Work autonomy		Severity and supportiveness in relations between fathers and sons
Hammond, 1954		Male's work dissatisfaction	Family tension
Bradburn and Caplovitz, 1965		Job tension	Marriage tension
Dyer, 1964		Male's job satisfaction	Affect on "whole tenor" of blue collarfamily life
Sennett and Cobb, 1973; Aronowitz, 1973; Rubin, 1976		"Dehumanization" in industrial work	Decreased possibilities for dignity and respect in the family

Notes to Table 3–1.

Sources of studies listed.

Arthur Kornhauser, "Mental Health and Factory Workers," *Human Organizations* 21, (1962): 43–46.

Alan A. McLean and Graham C. Taylor, *Mental Health in Industry* (New York: McGraw Hill, 1958).

Merrill T. Eaton, "Detecting Executive Stress in Time," Industrial Medicine and Surgery 36 (1967): 115–118.

Gilbert Steiner, "Day Care Centers: Hype of Hope?" in *Marriages and Families*, ed. Helena Lopata (New York: Van Nostrand, 1973) pp. 316–324.

R.L. Kahn and John R.P. French, "A Programmatic Approach to Studying the Industrial Environment and Mental Health," *Journal of Social Issues* 18, no. 3 (1962): 1–47.

Stanislav V. Kasl, "Work and Mental Health," *Work and the Quality of Life*, ed. James O'Toole (Cambridge, Massachusetts: M.I.T. Press, 1974) pp. 171–196.

James S. House, "The Effects of Occupational Stress on Physical Health," *Work and the Quality of Life*, ed. James O'Toole (Cambridge, Massachusetts: M.I.T. Press, 1974) pp. 145–170.

Bruce Margolis and William H. Kroes, "Work and the Health of Man," *Work and the Quality of Life*, ed. James O'Toole (Cambridge, Massachusetts: M.I.T. Press, 1974) pp. 133–144.

Erdman B. Palmore and Virginia Stone, "Predictors of Longevity: A Follow-up of the Aged in Chapel Hill," *The Gerontologist*, 13 (spring, 1973): 88–90.

Bruce Margolis and William H. Kroes, *Work and the Quality of Life*, pp. 133–144.

Donald Gilber McKinley, *Social Class and Family Life* (New York: Harper and Row, 1970).

S.B. Hammond, "Class and Family," in O.A. Oscer and S.B. Hammond (eds.) Social *Structure and Peronality in a City* (London: Routledge and Kegan-Paul, 1954).

Norman Bradburn and David Caplovitz, *Reports on Happiness* (Chicago: Aldine, 1965).

William G. Dyer, "A Comparison of Families of High and Low Job Satisfaction," *Marriage and Family Living* 18 (February, 1956) pp. 58–60.

Richard Sennett and Jonathan Cobb, *The Hidden Injuries of Class* (New York: Vintage Books, 1973).

Stanley Aronowitz, *False Promises: The Shape of American Working Class Consciousness* (New York: McGraw Hill, 1973).

Lillian Rubin, *Worlds of Pain: Life in the Working-Class Family* (New York: Basic Books, Inc., 1976).

of Americans—almost three out of every five—are 'satisfied' with their present jobs, and an additional 25 percent are 'very satisfied' . . ."[64] Further disagreement exists on how to measure job dissatisfaction.

Finally, the linkage between job satisfaction and psychological well-being is not clear. Psychological well-being appears to be composed of such elements as self-esteem, a clear sense of identity, goals, and values, and feelings of potency and efficacy. It is not clear, however, how much work satisfaction contributes to overall well-being (or conversely, how much work dissatisfaction detracts from overall well-being).

Melvin Seeman, for example, in an extensive study found little evidence that alienated work has the generalized consequences ordinarily imputed to it. Such generalization as he found was largely confined to the work realm. He suggests that workers simply come to terms with the only life they know and can reasonably expect for themselves.[65]

Several other studies suggest that because work is increasingly defined as less important in one's life than family life and leisure, work satisfaction may be contributing less to overall mental health. Earlier findings by Stanislav Kasl and by Leonard Sayles and George Strauss about the declining attachment to work roles[66] are supported by Daniel Yankelovitch's finding that

> When work and leisure are compared as sources of satisfaction in our surveys, only one out of five people (21 percent) state that work means more to them than leisure. The majority (60 percent) say that while they enjoy their work, it is not their major source of satisfaction. The other 19 percent are so exhausted by the demands work makes of them that they cannot conceive of it as even a minor source of satisfaction.[67]

Continuing Socialization at Work

As sites of the continuing socialization of adults, workplaces provide reinforcement of existing views and inspiration for new ideas, beliefs, and attitudes.

Studies of adult socialization to date have largely been undertaken in professional schools, rather than at the worksite. Some research suggests that the workplace is a major source of information on people's world view,[68] but we must await increased attention to the topic before drawing firm conclusions.

Learning about Gender Roles. In asking how the work world may shape the beliefs and attitudes that parents have about gender, we must recall the work-world characteristics outlined earlier in this section. Many workplaces, especially blue-collar workplaces, are either all-male or all-female environments. Gender learning here will be through contacts with others of the same sex. Many workplaces thus represent a continuation of the sex-segregation learning environments of adolescence.[69] The sex education appears to be adolescent in content as well: males discuss and joke about sex, often in exaggerated or bravado terms, and say little about intimacy or relationships; women continue patterns of talk that focus upon romance or on the relationships they have with their lovers or husbands, but with little discussion of sex.

In workplaces where both men and women can be found, structural aspects of the work world are also important. In most mixed-gender work situations, men occupy the higher prestige and higher paying jobs. Harris Schrank and John Riley[70] suggest that there are separate "gender pools" with jobs differentially rewarded (income) and ranked (prestige). A clear example of such differential ranking is offered in Rosabeth Kanter's *Men and Women of the Corporation.*

What is critical about this situation in terms of gender learning is that *most of a man's female colleagues will be subordinates and most of a woman's male colleagues will be superiors.* This is true whether they simply work in the same general environment (for example, in a school, in which most of the administrators are men and most of the teachers are women), or whether the women are connected vicariously to the men, as in secretary–boss relationships.[71]

Traditional patterns of gender relationship are likely to be supported by most work environments. Few people find themselves in equal positions with members of the other sex in their work lives. If they are equals, it is likely that either both are young and in low-paying jobs, or that the woman is older than her male colleague. Not simply in the society at large, but in specific workplaces as well parent-workers see that men have more control and more power than do women.

Even when people encounter others who defy traditional gender notions, they may not really learn anything and may not alter their general views. Kanter, for example, reports that many male business executives respond to career-women colleagues in terms developed at home:

> The existence of wives also had implications for women officially employed by the company. Just as the image of the secretary spilled over and infused expectations about other women workers, the image of the wife affected responses to career women at Indsco.

> Because corporate wives were generally seen to be content to operate behind the scenes and to be ambitious for their husbands rather than themselves and because of their use of social rather than intellectual skills in their hostess role, the image of women that emerged for some management men from knowing their own and other wives reinforced the view that career women were an anomaly, that they were unusual or could not really be ambitious, or that their talents must be primarily social and emotional rather than cognitive and managerial.[72]

Gender learning may also sometimes occur in the work situation because of the different expectations of men and women who occupy similar positions. Thus, for example, Schrank and Riley note that women in assistant positions often are not really in training for movement into the male job pool. Their jobs are characterized by emphasis on the helping role, little rewarding of initiative, performance assessment in terms of assistance offered, negative sanctioning of aggressiveness, and active constraint of management of others consistent with views of "what women are like." Because of (and reinforcing) gender stereotypes, women may also find they are automatically called upon for such tasks as coffee-making and note-taking at meetings.[73]

Sex and Eroticism. Little is known about the ways in which intimacy and sexual activity appear in the workplace. One researcher has argued:

> . . . as people who have interesting careers have always known, work is very sexy, and the people with whom one is working are the people who excite. A day spent launching a project or writing a paper or running a seminar is more likely to stimulate—intellectually and sexually—than an evening spent sharing TV or discussing the lawn problems or going over the kids' report cards.[74]

Worry may arise for some parent-workers about relationships developing at work that threaten family relationships. Launching projects, writing papers, and running seminars, however, are enterprises for a small proportion of the labor force. Furthermore, the unequal position of males and females probably precludes much of the supposed intimacy of workplace collaboration. People who have worked closely with other-sex colleagues have sometimes found such work highly competitive (as is much same-sex collaboration); others have developed deep, intimate, nonerotic relationships with colleagues; still others have developed intimate relationships that include sex.

The forms of sexual activity that arise in the workplace can be outlined, but—given the lack of data—we cannot indicate anything of their prevalence, except speculatively.

First, sexual allure or personal attractiveness serves as a prerequisite for obtaining some jobs and for being promoted in others. I am not referring here to such positions as actress or model, but rather to the importance sometimes attached to good looks rather than skill in the advancement of secretaries[75] and sometimes of junior executives.[76] In other instances, an employee's sexual allure may be used as a commodity by the employer—as in the case of restaurant owners who costume their waitresses to maximize their sex appeal to the customers.

A second instance of a sex-work linkage is the trading of sexual "favors" for employment or advancement. Sometimes sexual relations are initiated to obtain a job (the "casting couch" of the entertainment business). Again, because of the structural characteristics of the workplace, in most cases a male has the job (or raise or promotion) to give, and a female receives it as a result of the sexual liaison. There are probably few examples of men receiving raises or promotions because of their relationships with more highly placed women. The seduction, however, may take place in either direction. The director or executive or salesman may offer work rewards to the aspiring starlet or secretary if she will sleep with him; alternatively, she may initiate an affair with him in hopes of gaining better film roles, more office responsibility, or a higher salary.

Little concrete data exists on the prevalence of workplace affairs. Everyone seems to know of one, but actual cases may not be very common. For blue-collar men, the opportunity to meet women at work for possible sexual relations is limited by the gender segregation of their workplaces; in addition, opportunities diminish as the men grow older. Middle-class men have greater opportunities to meet women, and the opportunities will increase as the men age, receive promotions, and find more (and younger) women in subordinate positions.

For women the situation is more complex. The woman who has a paying job is likely to meet more men than if she were home full-time, but, as we have noted, many work situations are exclusively or almost exclusively female. As women move up in occupational rank, they are more likely to meet men at work. If they move very high (for example, to executive positions), they may find this lessens their appeal to some men (the reverse of the situation for men). They may also find that there are fewer men at their rank who are older—a problem, given the persistant belief that the male should be older than the female. Married men, therefore, are more likely than married women to have work-based sexual affairs, and such affairs are generally with younger, single women.

The importance of work in providing greater sexual opportunity outside marriage for both men and women (at least in the upper-middle-class) may not lie in the opportunity to meet colleagues of the other sex at work, but in the greater independence and control over time that work provides. Thus, both men and women in upper level work situations may have more possibilities for extramarital relations because they can create "open spaces" (long lunch hours, travel related to business, "after hours work") in which to consummate relationships developed in work or nonwork environments.

Finally, sexual harassment takes place in a number of jobs. Research on the forms of harassment and the extent of it in different work environments is in the early stages, but some harassment appears to arise in many work contexts. Again, it is largely women who are sexually harassed.

The Impacts on Family Life

The Search for Patterns. Workers differ in the respect that they gain from their jobs, in job satisfaction, and in their psychological well-being. In terms of jobs as learning environments, differences also arise in the beliefs, attitudes, and values related to interpersonal relations, to gender learning, and to sexual learning of those who people the workplace.

Such divergence may have led many readers to conjecture about effects on family life. The problem, however, is not whether there are effects (which everyone will agree), but whether there are *patterns of effect.* In the literature on work and leisure, the two alternative hypotheses on these patterns are generally referred to as the "spillover" and "compensatory" hypotheses.

A number of writers who have described sources of strain in work environments argue that these work strains create personal emotional strains that have a direct and strong impact on family life.[77] For example, a worker may come home from an exhausting, frustrating day at work and snap at "innocent" family members, or "take out work problems" on the family members. Conversely, elated by a solid compliment from the boss or a particularly rewarding work day, a worker may come home filled with joy and create a celebratory mood in the family. Thus, in the "spillover" hypothesis, a boring, pressured, alienating job will have a negative impact on family relations; a job that is interesting and unpressured will have a positive one.

In the "compensation" hypothesis, family life might be oppositely affected. After an exhausting, frustrating day, a worker may come home in great need of emotional support, full of warm feelings and appreciation for the family environment. In this case, negative work conditions contribute to more positive family relations. On the other hand, after exciting or rewarding work, the arrival home may bring a sense that family life is much blander, more routine, and boring. Here, negative work experience leads to positive family feelings, and positive work experience to negative ones.

These hypotheses can also be posited in terms of the consequences of working in environments that endorse different values. One might argue, for example, that executives—who work all day in environments that emphasize tough-mindedness, rationality and impersonality—will not be able to set aside these work values at the end of the day. Alternatively, however, one could argue that after working all day in such an environment, executives would welcome the opportunity provided by family life to express the warm, emotional, sensitive side of themselves.

If several possibilities exist, our problem is to collect sufficient data on the effects of work on the family in order to ascertain which particular effect is prevalent and under what conditions. We cannot simply say "work affects the family through the psychological response of family members"; we must inquire into the conditions under which family life is improved or suffers because of work life.

At this point such data are extremely scanty. Almost no research has examined (rather than making assumptions about) the connections between the two domains. Although there appears to be more support for the "spillover" hypothesis, many findings are unclear and often not statistically significant. It appears, therefore, that as yet we lack an understanding of the conditions under which spillover or compensation takes place.

Some researchers, for example, report job satisfaction as a major component of marital satisfaction—to the extent that it influences the whole tenor of family life.[78] Others find job satisfaction to be far more independent of marital satisfaction.[79] Melvin Kohn's and Daniel Miller and Guy Swanson's research on the effect of fathers' relationships to authority at work on their childrearing values and standards also yielded findings that support the spillover hypothesis.[80] But findings in both these studies are complex and not clearcut. The connections between the different world view of fathers in different occupations and their actual childrearing practices still warrant further investigation.

After summarizing these two studies and later replications of them, Kanter suggests yet other ways in which occupational norms and orientations may carry over to the family:

1. Through an emphasis on performance of discrete tasks as an indication of membership and contribution;
2. Measurement of performance by "objective" standards derived from more universalistic and impersonal notions of minimum and ideal performance (e.g., grades in school, frequency of intercourse, number of orgasms, amount on paychecks, and spotlessness of house);
3. Rewards and contributions controlled by each person given or withheld depending on performance;
4. Achievement (on discrete tasks) rather than intrinsic qualities as the measure of the person;
5. An expectation of "more" goods, services, rewards (comforts) with increasing seniority;
6. A legitimate ability to "fire" or "trade in" spouses if they do not meet standards.[81]

It must be remembered that these are suggestions rather than research findings. In addition, if such orientation should be found in the family, one would still need to demonstrate that they were "spillovers" from the work realm—rather than simultaneous developments stemming from a large common cause (that is, the norms of modern culture, exhibited in both the workplace and the family).

The available knowledge leaves us in some confusion, therefore, concerning the effect of the work experience on people as individuals. Likewise, we know little about the impact of work experience on the quality of spousing and parenting. Furthermore, as yet there is not agreement about what constitutes "good" family relations, and even where definitions are created of "marital satisfaction" or "marital happiness," researchers have difficulty measuring relevant factors.[82]

If the work environment affects how parents feel about themselves, and their knowledge and feelings about interpersonal relationships, gender, and sexual relations, then the work world may also have considerable impact on parents' teaching about a number of dimensions of sexuality, as Figure 3–1 suggests. What parents say and do with regard to gender roles, sexual values, life-style views and aspirations, sexual knowledge, expression of affection, body image, and sexual activity may all be affected by their work experiences. The nature of such effects remains to be investigated.

Gender Differences in Impact. Although many of the connections between the work world and family life remain unclear, we can anticipate different effects of work experiences for men and women—and hence for mothers and fathers in the same family.

First, it should be obvious from the foregoing discussion that the work world is generally a fairly conservative force in terms of gender-roles. We saw earlier that "getting out of the home" may be a liberating force for women in that it helps to create economic independence. Additionally, in the late 1970s labor force work, rather than housework, may also create a greater sense of self-worth for some women. Even where the work world gives women respect, however, it continues to offer them less respect than it does men. It is also less likely to serve as a source of emotional satisfaction and psychological well-being for women than for men.

Furthermore, the work world is not likely to be a place in which women pick up new notions about the appropriate balance of power between men and women. Few women will see at work alternatives to the male dominance that they probably experience at home. Thus, the work experience is not likely to give most women "lots of ideas" about radically changing the balance of power in their own families.

For these reasons, estimates of the major impact of mothers' work on children's socialization appear exaggerated. I would hypothesize that women who have paid employment will offer their children various models of women's work roles, but these women are not likely to act significantly differently (from wives not in the paid labor force) with their husbands—and thus are not likely to create examples of equal family partnerships (though they may be closer to equal than in nonworking mother homes). The impact of wives' work on children's gender learning, therefore should not be expected to be very great. Since most mothers are not likely to find that their work environments give them radically new information or attitudes on sexual questions, we should also not expect working to alter significantly the teaching mothers give with regard to sexual values or sexual activity. These changes are likely to come over time, and as opportunities and rewards in the labor force become more equitable for women.

V. CONCLUSION: DIRECTIONS FOR FUTURE RESEARCH

Throughout the paper there have been a number of statements about unknown linkages, missing data, and alternative interpretations. Let me now summarize what seem to be the major questions that remain, and the major areas in which research is essential.

1. Specification of Elements and Connections in the Model

While the data in Sections II through IV have allowed us to elaborate on the model given in the Introduction, it is obvious that there are many important gaps in our understanding. The problem is serious; in the absence of data, policymakers as well as researchers often rely on assumptions that, although they seem to accord with "common sense," may be unwarranted. It is assumed, for example, that if work leads to greater self-respect, family relations will be better. This may be true, however, only where family relations are already good; where relations are poor, greater self-respect generated by work roles may lead to termination of an unsatisfactory relationship. Parental separation may result from poor work experiences or from very good ones. The consequences for children are likely to be very different in the two instances.

In other cases, the absence of knowledge about one element makes it difficult to anticipate how the chain might be completed. For example, we do not know much about sexual experiences at the workplace (as discussed in Section IV). Nor do we know much about how such sex experiences in the workplace affect workers' sexual beliefs, attitudes, and behavior. Thus, it is difficult to assess how these experiences may eventually affect children.

Where data do exist, they often come from separate studies, and assumptions are then made about the connections between relevant findings. For example, there are now several studies of the effects of the employment of mothers on their children. Nancy Comer found that elementary school daughters of working mothers indicated enjoying a wider variety of activities in which males and females both engage; the daughters were also less sex-typed in their preferences. In addition, she found that there was greater admiration for their mothers by adolescent daughters of working women.[83] Grace Baruch found adolescent daughters of working women were more likely to attribute competence to women.[84] Tom Abernathy, Karl King, and Ann Chapman found that black adolescents with employed mothers were much less likely to view maternal employment as a threat to a stable marriage.[85]

Several interpretations of these findings may be postulated. The mother as a role model may be different; the employed mother may behave differently as a socializing agent or may say different things to her daughter about gender questions; the employed mother's different family role and way of relating to her husband may affect her daughter's notions. Another possible explanation would rely on the effect of the family's higher income, which permits access to a greater number of as well as the less traditional opportunities in the society. Yet another explanation might point to the fact that the employed mother spends less time with her daughter, supervises her less, and allows her more freedom. An accurate explanation of these findings would probably combine several of these factors. The point is that we do not know without conducting more systematic studies.

An example of a carefully designed study that begins to help our understanding of the process is one done by Marian Yarrow et al.[86] In a study of one hundred mothers (with intact marriages, of middle- and upper-middle-class status, with high school and college education) these researchers found a series of important differences in childrearing patterns between working and nonworking mothers. The fact of their working or not working did not affect what they did as mothers, but how they felt about what they were doing was critical in their performance. For instance, nonworking mothers who wanted to work—but who did not work because of a feeling of duty—felt that they had more discipline problems, had a poorer emotional relationship with their children, and were less adequate mothers. Working mothers with high school educations wanted more control over children and greater child participation in household labor, and they assigned more disciplinary tasks to the father than did nonworking mothers. Among college-educated working women, there was a greater tendency to plan additional compensatory activities with children.

The impact of labor force participation on childrearing is thus mediated by many other factors—including the mother's satisfaction with her own life. Better understanding of this process, however, awaits further investigation.

2. The Effects of Two Jobs, or Two Careers in the Family

As has been pointed out, most studies that have been done are studies of the effects of men's work or women's work. We do not know, however, whether two-job or two-career families experience additive or multiplicative effects. Thus, we need more studies that include parents in various combinations of work and family roles.

3. Parental Work Roles as Understood by Children

A further common assumption is that if we understand the character of the parent's work experience, we will thereby know what the child experiences. Parents, however, may not accurately describe what work is like, or if they do, the child may nonetheless misperceive it. A parent in a low-prestige job, for example, may describe it as honest work, or speak of the good colleagues on the job. The child may thus not know that there is anything "wrong" with the parent's job unless told so by others.

In a similar fashion, the child may conceive of the parent's job as something quite different from the parent's actual experience. The zookeeper and candy store owner may complain that the job has terrible hours, bad working conditions, and low income. Their young children, however, may find these jobs wonderful and may receive considerably more esteem from their friends than do the children of the business executive or automobile salesman. Consequently, we must do further investigations of how parents communicate about their jobs to children, and how children learn about the work world.

4. The Relative Impact of One's Own Family

In talking of the family and sexuality we have focused upon the impact of life in one's own family. Children, however, are affected not only by their own families, but also by the families of friends and neighbors. We must examine the relative impact of one's own family as opposed to other families with whom the child has contact. Many women who have entered the occupational world at high levels, for example, point to the influence of a neighbor or family friend who was employed and who showed that it was possible to have both a viable family life and a career. Other families may serve as models—either positively or negatively—of alternative gender roles, patterns of intimacy, and sexual relations. (This recognition, of course, is what leads parents to insist that their children have "nothing to do with" people with different patterns of sexual conduct—such as those in group marriages, those who engage in "promiscuous behavior," and the like. The very injunction, however, may make the other family an object of much greater interest.)

5. Studying the Assumptions and Consequences of Policy Changes

A number of policy matters have either direct or indirect implications for parental sexual teaching (for example, policy affects how many women work in the labor force, which, as we have seen, will affect sexual teaching to some degree). Often these matters have hidden assumptions, and their consequences are assumed rather than investigated.

One example is in the area of changing work schedules. Proposals for four-day work schedules have generally not met with success. Workers with families have found that while they themselves may have a three-day weekend, their spouses may have a traditional work schedule and their children will still be in school on a five-day basis. Workers in such cases cannot reorganize family patterns as easily as they may have anticipated. More success seems to be achieved, both for companies and for workers, with flexitime programs.[87] Because starting and finishing times are flexible, flexitime is often of great aid to individuals with other responsibilities and with difficult travel schedules.

Theoretically, flexitime has been developed to permit both men and women to adjust their schedules, and advocates often mention the benefits of shared responsibilities for husband and wife. At the present time, flexitime and other arrangements may, however, be more frequently adopted to permit mothers to perform the two roles of worker and mother, rather than to permit fathers to take a larger share of the child care responsibilities.

Flexitime could, of course, provide a way for men to spend more time with their children and engage in more of the tasks of home life. Whether it will be elected by them, and if the greater freedom will be used for family responsibilities rather than leisure remains to be ascertained.

NOTES TO CHAPTER 3

1. Rosabeth Moss Kanter, *Work and Family in the United States: A Critical Review and Agenda for Research and Policy* (New York: Russell Sage Foundation, 1977); Frank F. Furstenberg, Jr., "Work Experience and Family Life," in *Work and the Quality of Life*, ed. James O'Toole (Cambridge, Mass.: MIT Press, 1974), pp. 341–360.

2. Peter M. Blau and Otis D. Duncan, *The American Occupational Structure* (New York: Wiley, 1967).

3. Elizabeth Roberts, David Kline, and John H. Gagnon, *Family Life and Sexual Learning: A Study of the Role of Parents in the Sexual Learning of Children* (Cambridge, Mass.: Population Education, Inc., 1978).

4. Lynda Glennon and Richard Butsch, "The Distribution of Occupations on TV Family Series 1947–1977" (New Brunswick, N.J.: Douglass College, Rutgers University, 1978).

5. Gaye Tuckman, Arlene Caplan Daniels, and James Benét, eds., *Hearth and Home: Images of Women in the Mass Media* (New York: Oxford University Press, 1978).

6. This section has been adapted from a longer version in John H. Gagnon and Cathy Stein Greenblat, *Life Designs: Individuals, Marriages, and Families*, (Glenview, Ill.: Scott, Foresman and Co., 1978), Chapter 2.

7. Robert A. Lewis, "A Developmental Framework for the Analysis of Premarital Dyadic Formation," *Family Process* 11 (1972): 17–47.

8. R.R. Sears, E. Maccoby, and H. Levin, *Patterns of Child Rearing* (Evanston, Ill.: Row, Peterson, 1957); Lois W. Hoffman, "Early Childhood Experiences and Women's Achievement Motives," *Journal of Social Issues* 28 (1972): 129–155.

9. Jerome Kagan, "Acquisition and Significance of Sex Typing and Sex Role Identity," in *Review of Child Development Research*, eds. M.L. Hoffman and L.W. Hoffman (New York: Russell Sage Foundation, 1964), vol. 1, pp. 137–168.

10. J. Campbell, "Peer Relations in Childhood," in *Review of Child Development Research*, eds. M.L. Hoffman and L.W. Hoffman (New York: Russell Sage Foundation, 1964), vol. 1, pp. 289–322; Kagan, "Acquisition and Significance," pp. 137–168.

11. Roberts, Kline, and Gagnon, *Family Life.*

12. John H. Gagnon, *Human Sexualities* (Glenview, Ill.: Scott, Foresman and Co., 1977).

13. John H. Gagnon and William Simon, *Sexual Conduct: The Social Sources of Human Sexuality* (Chicago: Aldine, 1973).

14. Juanita Kreps and John Leaper, "Home Work, Market Work and the Allocation of Time," in *Women and the American Economy*, ed. Juanita Kreps (Englewood Cliffs, N.J.: Prentice-Hall, 1976), pp. 61–80.

15. Rosabeth Moss Kanter, *Men and Women of the Corporation* (New York: Basic Books, 1977).

16. John J. Gagnon, "Physical Strength, Once of Significance," *Impact of Science on Society* 21 (1971): 31–42.

17. Robert Blauner, "The Auto Worker and the Assembly Line: Alienation Intensified," in *Automation, Alienation, and Anomie*, ed. Simon Marson (New York: Harper & Row, 1970), p. 436.

18. Louise Kapp Howe, *Pink Collar Workers* (New York: Avon, 1977), p. 95.

19. Kanter, *Work and Family.*

20. Hanna Papanek, "Men, Women and Work: Reflections on the Two-Person Career," *American Journal of Sociology* 78, no. 4(1973): 852–872.

21. Gilbert Steiner, "Day Care Centers: Hype or Hope?" in *Marriages and Families*, ed. Helena Lopata (New York: Van Nostrand, 1973), pp. 316–324.

22. Yehudi Cohen, "'Househusbanding' or Suburban Mania," *New York Times*, 7 March 1976, p. 14.

23. Emmanuel Kay, "Middle Management," in *Work and the Quality of Life*, pp. 106–130.

24. Lee Rainwater, "Making the Good Life: Working-Class Family and Life-Style," in *Blue Collar Workers: A Symposium on Middle America*, ed. Sar A. Levitan (New York: McGraw-Hill, 1971).

25. Robert N. Rapoport, "The Male's Occupation in Relation to His Decision to Marry," *Acta Sociologica* 8 (1964): 68–82.

26. William F. Whyte, "A Slum Sex Code," *American Journal of Sociology*, July 1943, pp. 24–31; Lee Rainwater, *Family Design: Marital Sexuality, Family Size, and Contraception* (Chicago: Aldine, 1965); Furstenberg, "Work Experience."

27. Rainwater, *Family Design.*

28. Paul C. Glick, "A Demographer Looks at American Families," *Journal of Marriage and the Family* 37, no. 1 (1975): 15–26.

29. Karen S. Renne, "Correlates of Dissatisfaction in Marriage," *Journal of Marriage and the Family* 32 (1970): 54–67.

30. Rainwater, "Making the Good Life," p. 219.

31. William H. Chafe, "Looking Backward in Order to Look Forward," in *Women and the American Economy*, pp. 8–9.

32. Ibid., p. 17.

33. M. Geraldine Gage, "Economic Roles of Wives and Family Economic Development," *Journal of Marriage and the Family* 37 (1975): 121–128; Juanita Kreps, *Sex in the Marketplace: American Women at Work* (Baltimore: Johns Hopkins Press, 1971); Benson Rosen, "Sex Stereotyping in the Executive Suite," *Harvard Business Review* 52 (1974): 45–58.

34. For elaboration of this, see Gagnon and Greenblat, *Life Designs*, Chapter 12.

35. B. Bluestone, W.N. Murphy, and M. Stevenson, *Low Wages and the Working Poor*, Ann Arbor, paper for Institute of Labor and Industrial Relations, University of Michigan; M.H. Stevenson, "Women's Wages: The Cost of Being Female." (working paper, University of Massachusetts at Boston, 1972).

36. Kathryn E. Walker, "Household Work Time: Its Implication for Family Decisions," *Journal of Home Economics* October 1973: 7–11.

37. Joann Vanek, "Time Spent in Housework," *Scientific American* 231 (1974): 116–120.

38. J.N. Morgan, I. Sirageldin, and N. Baerwaldt, *Productive Americans: A Study of How Individuals Contribute to Economic Progress* (Ann Arbor: Institute for Social Research, University of Michigan, 1966), p. 102.

39. Walker, "Household Work Time."

40. J.P. Robinson and Philip Converse, *66 Basic Tables of Time Budget Data for the United States* (Ann Arbor: Institute for Social Research, University of Michigan, 1966).

41. Linda Lyle Holstrom, *The Two-Career Family* (Cambridge, Mass.: Schenkman, 1972).

42. Robert O. Blood, "Long-Range Causes and Consequences of the Employment of Married Women," *Journal of Marriage and Family Living* 27, no. 1 (1965): 43–47.

43. Lillian Rubin, *World of Pain: Life in the Working-Class Family* (New York: Basic Books, 1976), pp. 176–177.

44. Holstrom, *Two-Career Family.*

45. Susan Orden and Norman Bradburn, "Working Wives and Marriage Happiness," *American Journal of Sociology* 74 (1969): 391–407.

46. Lois Waldis Hoffman et al., *Working Mothers: An Evaluative Review of the Consequences for Wife, Husband and Child.* San Francisco: Jossey-Bass, 1974).

47. Ibid.

48. Kristin A. Moore and Isabel Sawhill, "Implications of Women's Employment for Home and Family Life," in *Women and the American Economy*, pp. 102–122.

49. Lee Rainwater, "Work, Well-Being, and Family Life," in *Work and the Quality of Life*, pp. 366–367.

50. Elliot Liebow, *Tally's Corner* (Boston: Little, Brown and Co., 1967), pp. 56–58.

51. Mirra Komarovsky, *Blue-Collar Marriage* (New York: Random House, 1964); Ruth S. Cavan, "Unemployment: Crises of the Common Man," *Journal of Marriage and the Family* 21 (1959): 139–146; Stanislav V. Kasl, "Work and Mental Health," in *Work and the Quality of Life*, pp. 171–196; Sidney Cobb and R. Rose, "Hypertension, Peptic Ulcer, and Diabetes in Air Traffic Controllers," *Journal of the American Medical Association* no. 4 (1963): 489–92.

52. Daniel Yankelovitch, "The New Psychological Contracts at Work," in *Work in America: The Decade Ahead*, eds. Clark Kerr and Jerome Rosow (New York: Van Nostrand Reinhold, 1978).

53. Rubin, *World of Pain*, pp. 174–176.

54. Janice N. Hedges, "Women Workers and Manpower Demands in the 1970's," *Monthly Labor Review* 93 (1970): 19–29; Kreps and Leaper, "Home Work."

55. Hedges, "Women Workers," p. 20.

56. Mart Witt and Patricia K. Naherny, *Women's Work—Up from 878: Report on the DOT Research Project*, (Madison, Wisconsin: Women's Education Resources, University of Wisconsin Extension, 1975).

57. Howe, *Pink Collar Workers*, pp. 237–239.

58. Studs Terkel, *Working* (New York: Pantheon, 1974).

59. James O'Toole, ed., *Work and the Quality of Life* (Cambridge, Mass.: MIT Press, 1974).

60. Blauner, "The Auto Worker," p. 436.

61. Howe, *Pink Collar Workers*, p. 35; Blauner, "The Auto Worker."

62. Mitchell Fein, "The Myth of Job Enrichment," *Humanist*, September-October 1973; Harold Wool, "What's Wrong with Work in America," *Monthly Labor Review*, March 1973; Irving Kristol, "Is the American Worker 'Alienated'?" *Wall Street Journal*, 18 January 1973; William Gomberg, "Job Satisfaction: Sorting Out the Nonsense," *American Federationist*, June 1973.

63. Terkel, *Working*; Rubin, *World of Pain*; O'Toole, *Work and Quality of Life.*

64. National Industrial Conference Board, "Most Americans Like Their Work," *A Topical Report* (New York: National Industrial Conference Board, 1978), p. 1.

65. Melvin Seeman, "On the Personal Consequences of Alienation in Work," *American Sociological Review* 32 (1967): 283–285.

66. Kasl, "Work and Mental Health; Leonard R. Sayles and George Strauss, *Human Behavior in Organizations* (Englewood Cliffs, N.J.: Prentice-Hall, 1966).

67. Yankelovitch, "New Psychological Contracts," p. 28.

68. P.J. Andrisani and G. Nestel, "Internal-External Control as Contributor to and Outcome of Work Experience," *Journal of Applied Psychology* 61 (1976): 156–165; M.K.L. Kohn and C. Schooler, "Occupational Experience and Psychological Functioning: An Assessment of Reciprocal Effects," *American Sociological Review*, 38 (1973): 97–118.

69. Gagnon and Greenblat, *Life Designs.*

70. Harris T. Schrank and John Riley, Jr., "Women in Work Organizations," in *Women and the American Economy*, pp. 82–101.

71. Ibid., pp. 90–91.

72. Kanter, *Men and Women of the Corporation*, p. 107.

73. Schrank and Riley, "Women in Work Organizations," pp. 94–95.

74. Robert Seidenberg, *Corporate Wives—Corporate Casualties?* (New York: Amacom, 1973).

75. Kanter, *Men and Women of the Corporation*, p. 76.

76. Herbert Marcuse, *One-Dimensional Man* (Boston: Beacon Press, 1964).

77. Liebow, *Tally's Corner*; Richard Sennett and Johnathan Cobb, *The Hidden Injuries of Class* (New York: Vintage Books, 1973); Rubin, *Worlds of Pain.*

78. William G. Dyer, "A Comparison of Families of High and Low Job Satisfaction," *Marriage and Family Living* 18 (1956), 58–60.

79. Robert C. Williamson, "Economic Factors in Marital Adjustment," *Marriage and Family Living* 14 (1952): 298–301; and "Socio-Economic Factors and Marital Adjustment in an Urban Setting," *American Sociological Review* 19 (1954): 213–216.

80. Melvin L. Kohn, "Social Class and Parental Values," *American Journal of Sociology* 64 (1959): 337–351; "Social Class and Parent-Child Relationships: An Interpretation," *American Journal of Sociology* 68 (1963): 471–480; and *Class and Conformity* (Homewood, Ill.: Dorsey Press, 1969); Daniel R. Miller and Guy E. Swanson, *The Changing American Parent* (New York: Wiley, 1958).

81. Kanter, *Work and Family*, pp. 42–46, 74.

82. Gagnon and Greenblat, *Life Designs*, Chapter 7.

83. Nancy A. Comer, "Working Mothers: How They Juggle Their Lives," *Mademoiselle* 81 (1975): 162–163.

84. Grace K. Baruch, "Maternal Influence Upon College Women's Attitudes Toward Women and Work," *Developmental Psychology* 6 (1975): 32–37.

85. Tom Abernathy, Karl King, and Ann H. Chapman, "Black Adolescents' Views of Maternal Employment as a Threat to the Marital Relationship: 1963–1973" (paper presented at the annual meeting of the Southern Sociological Society, Washington, D.C., 1975).

86. Marian Yarrow, Phyliss Scott, Louise de Leeuw, and Christine Heinig, "Child-Rearing in Families of Working and Non-Working Mothers," *Sociometry* 25 (1962): 122–140.

87. Kreps and Leaper, "Home Work."

 *Chapter 4*

Television as a Sphere of Influence on the Child's Learning about Sexuality

Hilde T. Himmelweit
*and Norma Bell**

INTRODUCTION

Franklin Fearing, writing about film, says:

Like the folktale, classic drama, primitive storytelling, or the medieval morality play, the film may be regarded as a means through which the individual understands himself, his social role and the values of his group. . . . He takes from the picture what is usable for him or what will function in his life. . . . Psychologically, an important aspect of this process is that of participation. It is the special characteristic of these media that the individual has an opportunity to project himself into situations and in some degree share in experiences otherwise denied to him. He may move into a world other than his own and acquire social identities and play social roles in many groups. . . . He may vicariously experience how other people react in a variety of situations.

The stimulus *value* of the motion picture is a question of the cultural values in our society which films express and the extent to which films communicate these values. . . . It is necessary to know something about the content of the film, the psychological needs of the persons who are exposed to it, the immediate setting, and the social and cultural forces operating on the persons who make the films and on the audiences for whom they are intended.[1]

*Hilde Himmelweit is recognized as a leading authority on television and social behavior. Having published some of the earliest studies in the field, she is coauthor of the book *Television and the Child.* Currently, Dr. Himmelweit is Professor of Sociology at the London School of Economics.

Norma Bell is a Research Associate at the London School of Economics.

We have quoted at some length Fearing's description of the ways in which films exert an influence on people because these ways apply equally to the influence of television. Although he does not label it as such, Fearing rightly adopts a "systems approach," in which he shows that a number of elements interact to produce given effects. Four such elements will be taken up in detail in this chapter. They are:

1. The content of the medium
2. The active participation of the consumer as selective processor of the events that impinge upon him
3. The situational factors that help modify or reinforce the impact
4. The fit between the content and the cultural values of the individual's environment.

CHARACTERISTICS OF THE TOPIC AND THE MEDIUM THAT FACILITATE SOCIAL LEARNING ABOUT SEXUALITY FROM TELEVISION

Social learning encompasses the acquisition of information, attitudes, values, norms, expectations, and skills, as well as the learning of appropriate role behavior and emotional states. A state of arousal, for instance, acquires meaning from the situation in which it occurs.[2] Attribution theory explains this phenomenon as the individual's search for meaning in interpreting experiences and events.[3]

Lawrence Kohlberg and Edward Zigler point out that the child's sex role identity, largely the result of self-categorization, is made early in life—at, or before, the age of three.[4] They further suggest that the judgment, "I really am and will always be a boy (girl)," is made during the regular course of development *relatively independently of the vicissitudes of social labeling and of reinforcers because the value of social reinforcers to the child is determined by his sex identity rather than the reverse.* The child's level of cognitive development, however, sets limits on the ways in which he or she interprets the environment with which he or she interacts and the effects of past interactions tend to modify perception of and reactions to the present.[5] While we agree that self-categorization occurs early in life, we can see no reason why the significance attached to it should not, as with other social identities, assume meaning through interaction with the environment. If this is so, general rules of social learning—modeling, imitation, observational learning, and reinforcement—would apply in the development of sex role identity.[6]

In the learning of sex roles, as of other roles, the immediate environment is the most important source. Not only does the immediate environment provide role models of "significant others" to whom the child stands in a close relationship, but it also permits rehearsal of a role, and (depending on outcome) the role's subsequent modification. Parents, siblings, peers, school, and neighborhood are important influences; the mass media, by comparison, are less important. Yet, as we shall show, there are particular circumstances in which television—and indeed, other media such as radio, books, and periodicals—have an important part to play in shaping the child's view of sex roles and sexual relations.

The degree of influence of the mass media varies according to the subject matter. When the content concerns a specific piece of knowledge (for example, how to hammer in nails) that can be taught by example, or taught explicitly by people with whom the child interacts, the mass media will add little. Where, on the other hand, the subject matter cannot be reduced to the presentation of facts but depends on observation, and where the subject matter is as amorphous and pervasive as that of sexual relations and sex roles (with much going on beneath the surface and behind closed doors), the mass media (as Fearing's quotation suggests), provide a very significant learning experience. The impact will be especially strong in situations in which direct questioning is rare, or in which both sides may be embarrassed and may lack an appropriate shared vocabulary.

Also, through television the child early on becomes aware that the relationships that he or she observes in the family may well be different from those that exist elsewhere. The child learns from TV because sexuality is a topic in which every child is interested from infancy. Finally, the child's search for information from television—as well as from all other sources—is the greater, the more he or she judges the subject (rightly or wrongly) to be taboo.

If the subject matter arouses heightened interest and, at the same time, there are barriers to obtaining the desired information, the influence of both peer groups and the mass media will be strong. Another factor has to do with the speed of social change. At a time of rapidly changing norms, when parents are less certain about the appropriateness of the rules by which they live their own lives, there is generally a greater search for relevant information. The birth control movement and the women's movement have led to new ideas, not only about abortion, but about sexual relations and performance. Sexual practices and preferences previously judged taboo are now openly discussed and sanctioned. The spate of sex manuals and sex therapy clinics would seem to indicate that it is now right and proper for individuals or couples to learn and become proficient in the skills of sexual play and intercourse. Many parents, nonetheless, are still apprehensive about talking about such subjects with their children or replying to questions when children read about the subject in newspapers or hear about it on radio or on television.

Television provides a strong learning experience not only because of the nature and content of the medium itself but also because of children's attitude toward and use of it. Below we list seven important factors.

1. Children spend a great deal of time watching television. A recent survey in Britain, based on 1,600 children of different ages, showed that children between the ages of seven and seventeen watch television between three and five hours a day (mostly during prime time) and view programs aimed at adults.[7] Takeo Furu's study carried out in Japan reveals that, on the average, children in the fourth through seventh grades watch four hours per day, and children attending the tenth grade watch three hours per day.[8]

   Recent figures in the United States indicate similar viewing habits and also suggest that viewing up to six and seven hours is not uncommon.[9] More time is spent watching television than on any other leisure activity.

2. Other things being equal, social learning occurs more readily the more the interaction is of the subject's own choosing—with people the subject likes, or in situations he or she enjoys. Television and the other popular media fulfill both these requirements.

3. Other things being equal, social learning through observation of models occurs more readily the more frequently the models are observed. Since all entertainment programs show how people behave—as actors, presenters, or interpreters—such programs all provide information about the behavior of men and women. The same is true of news and current affairs programs which show sex-appropriate behavior in different situations: at work, at demonstrations, at sports events, and at festivities. Occasionally, these latter programs provide information that bear directly on some aspect of sexuality, for example, Betty Ford's discussion of both her feelings about the removal of her breast, and the significance of the breast for women. The news also gives information about such sexuality-related subjects as marriage, separation, and divorce.

The majority of the programs that children watch are fictional programs—a few specifically designed for children, the majority aimed at adults. Fictional programs present an endless source of information about human relations, about the consequences of given behavior and attitudes, and about the emotional states that occur with or after given events: birth, death, marriage, divorce, young people leaving home, marital and family relationships, sexual relations, and the relation of sex to love.

In this context, the role of advertising must not be forgotten. The majority of advertisements dealing with house, beauty, and food products convey sex-related attitudes, values, and behavior. Research into the effectiveness of advertising for adults and for children leaves no doubt that—irrespective of whether or not the viewer believes the information provided is true—the viewer is influenced by what is shown.

4. Other things being equal, the influence of the message is the greater the more there is an underlying consistency in the information conveyed—provided there is some variation around the basic theme to facilitate the isolation of a general rule or principle. This is particularly the case with regard to sex-related roles portrayed on television. It should be noted here that effects rarely depend on any one individual program—although this does happen occasionally because of a program's excellence, because of a particularly moving incident, or because of the close fit between the program's theme and the individual's needs. But most effects occur according to the "drip" principle—as a result of messages received not from one, but from many programs.[10]

5. Other things being equal, children will model themselves more on characters with whom they can readily identify. An important feature here is the setting in which the action takes place. Himmelweit et al. found that children were more afraid when a murder occurred in a kitchen or bathroom than when it occurred in some more unusual surroundings—for instance, a historical or science fiction setting.[11] *The closer to home it is, and the more it is perceived as real, the greater the impact.*

6. Other things being equal, the influence of the message will be the greater (1) the more the source is liked, (2) the more the message is conveyed in a setting that creates emotional involvement, and (3) the more the viewer's attention is distracted from the message itself.[12]

7. Other things being equal, the influence will be the stronger the more the behavior of the model is rewarded, and will be weaker when the behavior is ignored or even punished. Albert Bandura and Richard Walters have demonstrated the operation of this principle in their studies of children's imitation of aggressive behavior.[13] In tests on two groups of nine- to eleven-year-olds, the children were assigned to either a "reality" condition (where they were told they would be seeing a television newsreel of a student riot) or to a "fantasy" condition (where the same film was used but the children were told that it was a film about a student riot). Comparing the aggressive behavior of the two groups after seeing the film, the researchers found that the "reality" condition stimulated aggression, while the "fantasy" condition reduced it. They suggested that the classification of a program as fantasy or reality is dependent on stimulus cues, direct labeling, and individual characteristics of the perceiver.

In summary, the more the child voluntarily chooses to use a mass medium (or the more the medium is enjoyed), the more the medium provides information in a manner that involves the viewer, the more it provides a consistent set of messages whose consistency is enhanced through small variations around the same basic theme, and the more such messages lead to reward (that is, are positively reinforced), the more likely it is that the medium will exert an important influence. Given the children's interest in sexuality and the barriers to acquiring sufficient information from the immediate environment, it follows that in the area of sexuality, television acts as an important source of influence.

The validity of the statements above is supported by a substantial body of research. The effectiveness of a stream of consistent messages has been borne out by studies investigating violence and aggression, and recently from studies examining the influence of television on prosocial behavior.[13] Although such studies have largely been correlational, they have been supported by the results of experimental studies.[14] The influence of television's messages on the behavior of young people may occur (1) through the acquisition of new responses, (2) through altering the likelihood of the performance of previously learned responses either because of a changed expectation about the rewards or lowered or raised inhibitions regarding their performance, or (3) by altering the meaning attached to that performance. Bradley Greenberg and his coworkers have studied the influence of television violence on perceptions.[15] They found that after viewing violence on television, children were more ready to see violence as a legitimate way of resolving conflict, and to see it as more prevalent and as occurring more frequently in their immediate environments.

George Gerbner found the tendency to treat information obtained from fictional programs as a source of information about the "real world" occurs among adults as well as among children.[16] Heavy viewers, who are more exposed to violence on television, consistently put higher estimates, on the likelihood that they will be involved in a violent incident, than light viewers do. This difference holds up for viewers of various ages, both sexes, and with and without college education. The technique used was interesting. Viewers were provided with a whole range of statistical estimates regarding the likelihood of encountering violence, the percentage of crimes that are violent, and the likelihood of men employed in law enforcement and detection to be violent. The estimates of heavy viewers of television drama, more often than light viewers, came closer to the frequency of the event on television than to the frequency in real life. The frequency of witnessing violence vicariously, through the viewing of fictional programs, provides (as Fearing suggests) a yardstick for the assessment of real life events.

In a different example of incidental learning, Ann Beuf found that the more children view TV, the more they sex-type certain occupations.[17] There can be little doubt that the frequency of exposure and hence the frequency with which the consistent messages are received are an important factor in influencing learning and imitation.

THE CONSUMER AS AN ACTIVE PARTICIPANT IN VIEWING

The consumer is an active participant in events; he or she selectively processes events. This applies to television as well as to happenings in the immediate environment. Below we list factors that affect this interactive process.

1. Other things being equal, the child will model himself or herself on persons of the same, rather than of the other sex; modeling takes place on characters who are perceived as similar to the self. This statement is only partially true since, as we show below, another influence modifies the operation of this principle for girls.

   Eleanor Maccoby, William Wilson, and Roger Burton found that when they showed a film in which the two central characters were a man and a woman, girls spent more time than boys watching the woman.[18] The amount of time spent watching the man, however, was the same for both sexes. When children were asked to recall what the characters did, the differential recall of same-sex characters was confirmed for all but two types of content: male aggression and romantic love scenes.

2. Other things being equal, imitation occurs more readily if the model has one or more of the following characteristics: attractiveness, legitimacy, expertness, coerciveness, and the ability to give rewards.[19] Four of these five characteristics would seem to favor imitation of the male model. Paul Mussen and Luther Distler found that boys who, in doll play, showed a strong preference for father figures, saw their own fathers as powerful; on the other hand, girls, compared with boys, were more ready to imitate the behavior exhibited by a model of the other sex.[20] "This difference probably reflects . . . the relatively greater positive reinforcement of masculine-role behavior in our society."[21]

Marked sex differences can be seen in the way children use mass media material.[22] Studies indicate sex-typing in what is learned: girls, for instance, show more incidental learning from a film depicting a domestic situation[23] and more identification with the heroine than do boys.[24]

Using an ingenious picture game, Beuf asked four- to five-year-old children to indicate whether certain situations shown in the pictures were or were not as they should be.[25] While 86 percent thought it was appropriate for the father to feed the baby, 51 percent considered it not appropriate for the telephone repairperson to be a woman. At this early age, children already see the world divided into male and female roles. Both see males' jobs as more attractive. Many girls preferred tasks carried out by boys or men, while boys showed a great reluctance even to consider what they would do if they were girls; the idea was anathema to them.

Recognition of sex-appropriate behavior thus occurs at a very early age. By the age of three, children have many accurate perceptions of sex-role behavior and can correctly identify which sex would use certain sex-typed objects. While from an early age children imitate same-sex models, the exclusiveness of such modeling differs between boys and girls. Since fewer women are shown on television (and those who are depicted as less powerful and less expert), M. Mark Miller and Byron Reeves showed that girls have fewer role models of their own sex among television characters.[26] When 200 children in grades three to six completed questionnaires in which they were asked to name any people on television that they would like to be like when they grew up, boys nominated significantly more models than girls did ($p < .05$)—all of them male. Twenty-seven percent of the girls chose male characters to imitate.

3. Other things being equal, the impact of television will be greater the more it is used and valued, and will also be greater where there are few alternative, accessible sources of information and entertainment. The word "accessible" is used advisedly, since for some people reading is often not an accessible alternative.

   Working-class children compared with middle-class children, blacks compared with whites, and the less educated compared with the more educated in any given age group tend to view more television.[27] Children who view more TV are also more likely than those who view sparingly to imitate characters they see and to be more uncritical of what is offered.[28] For them, the medium provides a way of seeing what goes on in a world to which they would otherwise have little access.[29]

4. The environment in which the child is brought up can be rich or limited in alternative sources of information. The immediate environment of a small village, compared to that of a large city, provides a very restricted range of role models. In the village the implicit messages about different roles televised entertainment provides are therefore more likely to influence the child's notions of appropriate behavior in novel situations.

   An early study by Himmelweit et al. provides evidence on this point.[30] The study was carried out at a time when it was still possible to match each child who had a television at home with a child from the same classroom and of the same sex, social class, and intelligence who had neither a television at home nor the opportunity to view one as a guest. The children were asked to describe the living room of an ordinary family and that of a rich family. There were no differences between viewer children and control children in the descriptions of the living room of an ordinary family. Both groups described the living rooms in their own homes—they drew on firsthand experience. In the description of the living room of a rich family, however, differences did arise. TV viewers significantly more often placed in the living room of the rich those hallmarks of wealth that are usually found in television plays—cocktail cabinets, chandeliers, and so on. The children here were drawing on the experience gained through television.

Of more relevance to the current discussion were the differences found when the children were asked to name well-paying jobs. Children without television more often named the well-paying jobs about which they had heard. (In one town, these jobs were in a shoe factory.) The viewers, on the other hand, had a much wider range of jobs to draw on—such as judges, lawyers, doctors, and businessmen—all jobs that had been seen on television. Television here had provided a canvas of experience that the children treated as real even though most of the information came from fictional programs.

This evidence is relevant in that it shows that such information need not be "taught"; no one on television explicitly informed the children that some jobs were better paid than others. Instead, they inferred this fact from a whole variety of cues that the programs provided, including the deference that interviewers show to people in prestigious occupations. The process whereby children pick up cues about appropriate sex-linked behavior and norms is subtle. In the Beuf study, the more a child viewed, the more likely he or she was to apply sex stereotypes to careers.[31]

5. Other things being equal, messages that relate to the needs of individual viewers will make more of an impact than those that are more remote. Where the message has no relevance it will not be attended to. The findings reported above showed a difference between viewer children and control children in the naming of well-paid jobs only among the older children—the thirteen- to fourteen-year-olds who would be shortly entering the job market. The ten- to eleven-year-olds were not sufficiently interested to attend to the implicit message.

6. Television will make more of an impact the less formed the attitude or behavior in question. In the Himmelweit study the attitudes toward foreigners of viewer children and control children were compared, and a decrease in ethnocentrism was found only among the younger children (ten- to eleven-year-olds).[32] The attitudes of the older children were already too well established for television to make a significant difference—at least in the short run.

This is not to say that a steady bombardment of consistent message cannot, or does not, make inroads into attitudes or beliefs that on first inspection would seem to be well established. Indeed, the ease with which television can affect perceptions is remarkable. Using a before-and-after design, Donald Roberts examined the effect on fourth-, fifth-, and sixth-grade California school children of a television series called *The Earth a Big Blue Marble* in which the lives of children from other countries were shown.[33] Attitudes after the series, compared with those before, showed that the children had learned to see more similarities between U.S. children and children from other countries and were less likely to believe either that U.S. children were better off, or that most children would prefer to live in the United States. As in the Himmelweit study, however, the decrease in ethnocentrism in the younger children was greater than the decrease among older children.

In the field of sexuality, too, we would expect that television would exert a strong impact. Children, though aware of their own sexual identity at an early age, have little direct experience with a variety of adult male and female models. Television provides them with many models. The ingredients for a strong impact are: an interest in the subject matter, an absence of well-established attitudes, the possibility of identifying with models of the same sex through fictional programs, and the perception of such fiction as real. Such factors apply particularly to the young and (within each age group) to those with fewest alternative role models or least access to alternative information.

THE RELATION OF PROGRAM CONTENT AND CULTURAL VALUES

In children's programs, women (compared with men) are more affectionate, submissive, and fragile, and less ambitious, self-confident, individualistic, and dominant.[34] Even a program as carefully prepared as *Sesame Street* has much sexual stereotyping. Women in this series, as elsewhere on television, are under-represented. Those who do appear are passive and unadventurous, shown in less interesting jobs, and generally bound to the home.

According to Carolyn Cathey-Calvert,

> "*Sesame Street* has sufficient cultural introspection to enable it to eliminate the portrayal of ethnic sterotypes, but it lacks this introspection when portraying females, and there, it continues to portray sexual stereotypes."[35]

Cathey-Calvert points out that if we are to have equality among people, then it must be among all people—not just among the males. The fact that *Sesame Street* is full of sexual sterotyping—even though the show has effectively eliminated racial sterotypes, and despite the fact that it has more women producers than other television shows—is, in itself, an interesting reflection on the degree to which sexual sterotyping is woven into the fabric of the society. Reversing this sterotypic image in future programs will require a conscious effort on the part of the producers.

Michele Long and Rita Simon examined the role and status of women on children's TV programs by doing a content analysis of twenty-two programs aired on Saturday morning and in late afternoon hours.[36] The researchers coded the amount of time that any woman appeared and her status, interactions with males, and physical characteristics. They concluded that during the years of the study, 1970–72, the general image of women changed little. The women tended not to work in jobs outside of the home; they were generally married; if single or widowed, they are mainly concerned with "getting their man." As a rule, the women were young and attractive and much concerned with their appearance. They tended to be overemotional and dependent on husband or boyfriend. The few who were shown at work were in low-level occupations.

In the Linda Busby analysis of commercial network programs directed to children, the coders used the semantic differential format to examine a range of personality characteristics and found that twenty-four of the forty programs differentiated significantly between the men and women who appeared in their shows. The men were more ambitious, competitive, adventurous, knowledgeable,[37] independent, aggressive, dominant, logical, and self-reliant. Women were less violent and brave and more affectionate, romantic, emotional, submissive, and timid. Compared with men, women were followers rather than leaders. Busby also found that males dominated in the home as well as in society in general. Men occupied not only more prestigious occupations, but also a wider range; males were shown in forty-two different occupations, while women were shown in only nine. Women generally fit into the traditionally accepted roles of the wife, mother, or girlfriend who remains in the home, while the men work outside. Girls were engaged in traditional girls' roles of baking and sewing, while boys were in boys' roles—mostly in sports. Extending the analysis to cartoons, Busby found similar male-female stereotyping.

In action programs, which the young view a great deal, women—if they are featured at all—are either helpmates (as in Westerns) or inarticulate ornaments or impediments (as in crime programs).[38] For instance, in crime programs, the woman is viewed as an impediment if she demands attention or if the criminal falls in love with her (since this causes him to temporarily lose his interest in his vocation of being a criminal). Any assertion of individuality on the part of the moll is punished by a blow, or, if she wants to free herself, she is generally killed off by henchmen who neither hesitate nor express regret. Crime programs epitomize the role of the woman as a chattel to be disposed of when it becomes irrelevant or, worse, a burden.

By sheer numbers, men dominate the scene.[39] They are 75 percent of all major characters. In adventure shows they are 85 percent of the major characters. Even in situation comedy, they constitute a majority of 55 percent. The personalities of men are generally more complex than those of women: of the males 46 percent have an indeterminate marital status, and 53 percent have indeterminate parental status. The percentages of women of indeterminate marital and parental status are 11 percent and 19 percent respectively.

The relationships involving women tend to be those between men and women and between mother and daughter; if the relationship is between two women, it is generally concerned with marriage and betrayal.[40] Rarely are relationships between two women shown as affectionate and warm. Where both husband and wife work, conflict arises. Marital and family relationships constitute 41 percent of women's interaction but only 18 percent of men's interaction.

Natan Katzman, in a study of television soap operas, found that characters were evenly divided by sex, but that this equal distribution disappeared when the characters were further classified by age, occupation, and marital status.[41] More women were portrayed as young adults and more men as mature. All children in the sample were males. Where marital status could be determined, more men had never been married and more women were widows. For the most part, however, men and women remained unmarried.

Other research has shown that in close relationships portrayed in TV programs where men and women have similar personalities and perform similar tasks, there is likelihood of conflict and violence. In the portrayal of a relationship where the woman is nurturant and the man is independent, the relationship is more likely to be peaceful than if both partners are independent.

Overt kissing, embracing, or affectionate touching appear significantly more often in situation comedies than in crime, adventure, or dramatic programs.[42] Indeed, there is a striking lack of intimacy in dramatic programs. Although heroes and heroines are portrayed as leading exciting and rewarding professional lives, they appear to endure austere private lives that lack physical or verbal expressions of tenderness.

In recent years some upgrading of the role of women has taken place. Comparing one week's content analysis of prime time television programs across the span of four years (1971 to 1974), John

Seggar found an increase in the percentage of women characters.[43] The ratio of four men to one woman in 1971 had changed to two to one by 1974. In programs such as *The Man From Atlantis*, *Star Trek*, *Bionic Woman*, and *Charlie's Angels*, women do difficult and daring jobs. However, *Charlie's Angels* illustrates well how the traditional roles are dished up in new guises. The women investigators are modern day Mata Haris. Women police officers (as in *Police Woman*) are shown to lose their cool more readily than their male colleagues, to get more emotionally involved, and to evoke more tender concern for their safety. In a subtle way, they manage to repeat the sexual stereotypes shown in soap operas.

Such stereotyping is especially insidious since (as research has shown), the influence of a role model depends upon the extent to which it is rewarded, and in television the traditional role is rewarded. In an atypical role, such as the successful working woman, the woman tends not to be rewarded; indeed, she is punished for occupying such a role by having problems in her personal relations with lovers, husband, or child. Women are significantly more often portrayed as either very rich or very poor—with wealth generally achieved by marriage or family background, rather than by work. Yet on the few occasions where wealth is achieved by work, it tends to be gained at the expense of happiness. This is not so with men.

Although most of the researchers have concentrated on the woman, sexual stereotyping is also evident in the portrayal of male characters. Nearly always in control, coolly planning, emotionally uninvolved, the man is rarely seen caring for family or home. The important world is outside, even though he comes home to wife and a meal. He rarely has much of a personal life, and where there is one, it has to take second place to the requirements of the job. When he is shown in the role of a father, he is rather ineffectual and weak, manipulated by adolescents and mother alike.

The sex stereotyping in commercials is even stronger than that in the shows themselves. Most commercials show women employed in domestic activities. We are led to believe that having done the housework or cooked a good meal, the woman will be rewarded by a kiss or a smile from her man. In general, men appearing in commercials do not clean, prepare meals, or shop (except for car accessories).[44] The emphasis in commercials is on youth and physical attractiveness; the most frequent behavior shown is women's diligent labor, harrassment, and vacuous astonishment.

When it comes to an explanation of the nature of the product, however, men take over—in the form of male presenters or male voices. This division of labor re-emphasizes that expertise, authority, and knowledge are male prerogatives. Ellen Wartella and J.S. Ettema found that children are sensitive to auditory cues in commercials,[45] and Stephen Kline, in an experimental study, showed that information conveyed by voice-over is absorbed even though the combination of auditory and visual information is more powerful than either alone.[46] (Auditory information makes an impact because, unlike visual information, it requires no encoding before comprehension or storage can take place.) The parallel between the commercials and *Charlie's Angels* is particularly striking. In *Charlies's Angels*, women do the daring work while Charlie's voice, over the telephone, assigns the tasks and assesses performance.

The intensity of the sexual stereotyping and the repetitive nature of television advertising, plus the pairing of attractive incomes with advertised products, have led to the suggestion that advertising can almost be likened to instrumental conditioning.[47] Indeed, Scott Ward and his associates have shown the impact that television advertising has on cognitions and beliefs.[48] Although that study tended to focus on the product, a great deal of other learning material is presented in commercials that, because it is associated with the desired product, will be positively viewed.

The types of television content analysis just described go well beyond head-counting, or the simple consideration of the occupations of men and women, to an analysis of the nature of interpersonal relations and an assessment of the personality traits of positive and negative male and female role models. The conclusions are clear and well-supported. All studies—regardless of coders, samples, or methodologies used—testify that sex stereotyping is frequent and pervasive.[49]

EDUCATIONAL TELEVISION PROGRAMS

Here we turn to the effectiveness of special education programs, which are, of course, viewed in special ways—in the classroom or at home as grist for the next day's classroom discussion.

A wide variety of educational television programs exist. Some are designed to impart specific pieces of knowledge about sexual functions. Others not only aim at teaching the child about the body and the way conception and birth occur, but also place equal emphasis on making the child aware of attitudinal and emotional aspects of sexuality. That is, they aim at imparting knowledge and influencing attitudes.

Recently in England, two series designed to provide facts and create a certain atmosphere around sexuality were shown on primary school television. One of the series, was part of a longer nature series being shown weekly during school hours.[50] In this series, a live birth was filmed and the correct Latinate names of sexual parts were given, along with an explanation of the parts. The schools provided very good test facilities, thereby making it possible for observers to sit in the classroom, note the general atmosphere, the questions children asked, and the teachers' replies. Each child completed a questionnaire at the end of the series, and after a three-month interval. Parents and teachers were asked to express their own attitudes about the programs and were invited to record the reactions the children displayed.

Interestingly, the effects of the program did not vary according to parental attitudes towards sex or feelings about the appropriateness of sex education in schools. Of the 200 eight-to-twelve-year-old children studied, all learned a good deal. Not surprisingly, those with least previous knowledge learned the most. There were no social class differences, and ability differences mattered remarkably little.

In addition to this increase in knowledge, the children's attitudes toward childbirth and nudity changed. They came to be less anxious about the pain of childbirth and the "naughtiness" of nudity. At that age, many children consider it "naughty" to look at someone in the nude or to show themselves without clothes. Interestingly, there was no correlation between change in attitude and information acquired. It would appear that two separate learning processes occur, a cognitive one and an affective one, both of which were shown to persist over a three-month period.

We have described the study in some detail, not only because we were closely involved in it, but because for the first time it was possible to use suitable pictorial and questionnaire material in order to elicit information from relatively young children about their attitudes toward sexuality and bodily functions.

The second program, *Living and Growing*, produced by Grampian Television, also proved successful.[51] Here, however, data was obtained solely from teachers who, particularly on topics of this kind, are not necessarily accurate judges of pupils' attitudes and knowledge. Bradley Greenberg and Byron Reeves found that children's attribution of reality to television portrayals was strongly related to the child's perception of the views of friends and family and to intelligence, school grade, and amount of television viewing.[52]

The impact of educational programs depends on the readiness and interest of the teacher as much as on the program itself, but there is little doubt that both attitude and knowledge can be taught. Nor is it surprising that such programs can be effective. Several studies have demonstrated that prosocial behavior can be influenced by programs specifically designed to do just that. Summarizing the experimental studies, George Comstock points out that not only attitudes but also the matching behavior can be influenced—the display of affection and generosity.[53] What makes these studies relevant to the discussion of sexuality is that the emphasis of the programs is on altering the value that the young viewers place on the acts viewed. In a study carried out in our department, Philippe Rushton used a filmed model to induce altruistic behavior in primary school children (giving away tokens they had won in order to help poor children.)[54] He further showed that the readiness to give persisted over a period of months. Although he could induce such behavior through a filmed model, the film was (as one would expect) less effective than a live adult model who gave away his or her own tokens. This particular study is important because it tested the effectiveness of advocacy with, and without, the appropriate accompanying behavior. Advocacy alone was less effective than behavior alone, while advocacy and example made the largest impact.

THE VIEWING SITUATION

The importance of the viewing situation—that is, the presence of others, generally adults, and subsequent discussion of the program—has been studied in a number of settings, though not specifically with regard to sexuality. With young children, in particular, the presence of the mother enhances the enjoyment of the program, provides a cue that the program is interesting (since the mother watches it), and increases gain in knowledge and, when the behavior shown is praised by the mother, the imitation of such behavior.

Fearing, Furu, and others have stressed the importance of the viewing situation in affecting the degree of learning or imitation that can occur.[55] Steven Chaffee and Jack McLeod have shown that, among adolescents, political information given on television was more readily absorbed and recalled when the issues were discussed in the home.[56]

Jerome Singer and Dorothy Singer carried out an experimental study in which they randomly assigned three- to four-and-a-half-year-old children to one of four groups.[57] One group watched the program *Mister Rogers' Neighborhood* for half an hour each day for two weeks. A second group watched the same program, but with an adult present. A third group watched no television but spent the same amount of time playing with an adult. A fourth group acted as control. Significantly, more children who watched the program with an adult imitated the prosocial behavior of Mr. Rogers than did those who viewed on their own.

Using Israeli children, Gavriel Salomon compared the effect of *Sesame Street* on the learning of cognitive skills of two matched groups, one of which watched on their own, while the other watched with their mothers.[58] In the first group, there were large social class learning differences, with middle-class children gaining more from the program than working-class children did. In the group where the mother was present, these differences largely disappeared; the working-class child seemed to learn as well as the middle-class child.

We also studied the effect of reinforcement in an experimental study for which special film was prepared.[59] Prior to seeing the film, school children were asked to indicate how much they liked two animals, an elephant and a rhinoceros. They were told that they would later have to evaluate the film. The film showed a boy of the same age as the children tested, looking in turn at pictures of the two animals. Each time the elephant appeared, the model received a shock—his hand on the armrest of the chair jerked and he grimaced. When the rhinoceros was shown, the model's hand and face relaxed. Afterwards the children were asked to say how much they liked the film and to answer various questions, including ones about the two animals. A significant decrease in liking for the elephant (the animal associated with negative reinforcement) occurred; the change in attitude was still found to operate some two months later.

It must be remembered that the children were unaware that the shock was administered in relation to the showing of the pictures of the animals. A more dramatic version was also tested.[60] While this second version caused greater immediate anxiety on the part of the children, vicarious conditioning did not take place. The essential contingencies had been lost. Here is an example where the mode of presentation makes a substantial difference and where the less dramatic of the two versions caused the greater and more lasting impact. Other studies we have carried out similarly point to the importance of presentation and of determining very precisely the concordance between the visual and the verbal information conveyed.[61]

In summary, we feel that television is useful for providing information about those areas of life that people in general find embarrassing or difficult to discuss openly. The validity of this statement is substantiated by the recent sex education programs in schools, limited though they may be. Also, we conclude from the studies mentioned above that the rather sweeping generalizations sometimes made about the impossibility of either imparting knowledge about sexuality or changing attitudes toward it are unwarranted. It seems that much depends on the quality of the programs. Also, experience indicates that the effectiveness of the programs as teaching instruments can be enhanced by subsequent discussion either in the home or in the school.

SUMMARY AND IMPLICATIONS

We have reviewed here relevant studies and placed them within a conceptual framework that analyzes the conditions under which the mass media are likely to have an impact. In the case of sex roles and sex—appropriate behavior, we have shown that television offers a very stereotyped picture and that, if anything, television lags behind rather than leading in the move toward greater sexual equality. Part of this lagging behind is the tendency of the medium not only to retain old stereotypes but also to give them new occupational guises (the police officer who looks like a fashion model, or the female computer expert who risks all for her man).

Evidence has been presented to show why, in the sphere of sexual development, television and film are likely to provide important learning about sex roles and providing information about sexual relations. This is not only because of the amount of time children spend watching these media or the degree of stereotyping in their content, but is also due to the nature of the topic itself.

The age at which a mass medium makes the greatest impact varies with the topic. We suggest that both the immediate environment and the reinforcement offered by a medium's stereotypes lead to such views being relatively fixed by middle childhood. Information offered subsequently, for instance during adolescence, when the young make plans for their occupational futures, is therefore less likely to make much of an impact.

Television has failed to respond to social change in society. In the United States, the majority of women now have a job. They work in many fields and, in increasing numbers, hold highly responsible positions. Many of these working women are also raising children. In their new dual role, they face new problems that demand new solutions. But television does not reflect this reality. Instead, the symbolic reality that it offers is old-fashioned, unequivocal sexism. Why is there this out-of-date stereotyping? It has much to do with the structure and financing of the broadcasting industry, which fears deviation from established norms. To be out-of-date is less risky than to alienate a portion of the viewing population through an image of men and women that the viewers might not share. The cause may also lie in the fact that the industry itself has few women in decisionmaking roles. But even where women do act in that capacity, as in the case of *Sesame Street*, stereotyping of sex roles has continued. It may well be that attitudes are so deeply ingrained that the industry is insufficiently aware of its portrayal of the sexes.

What can be done about it? First, we believe more research is needed into the way children develop ego ideals and ideas about appropriate behavior. Second, groups like Action for Children's Television, Parent/Teacher Association, and other associations should continue to take definite action by monitoring what is presented and subsequently attempting to influence what is shown. The recent attempt to reduce the amount of TV violence by publishing the names of firms that sponsored the more violent programs was remarkably successful. It shows that, without infringing on the First Amendment, influence over content can be exerted. Since commercials are particularly guilty of excessive and obvious stereotyping, concerned groups should have no problem publishing the names of products with offensive commercials.

We are not suggesting a code of practice. That would be as ineffective as the whiter-than-white code of practice about violence to which most television companies subscribe. Instead, we would advocate that ways be found to make the decisionmakers in the industry, the writers and producers, far more aware of what is in fact presented, to acquaint them with the results of monitoring. Further, there is a need to make the industry better informed about the views of the public, so that it can reflect rather than lag behind the public's outlook.

Periodic content analyses should be carried out. These should include an analysis of the language used and of the attitudes and values conveyed, rather than simply a head count of numbers of men and women and the classification of occupations they are shown to follow. It would also appear to be necessary to make teachers aware of the strength of such stereotypes and to provide mass media material that might help them to broaden their pupils' perspectives—and to do so at an age when the children's attitudes are not yet firmly fixed.

It is hoped that the industry, perhaps with the help of foundations, might create programs specifically designed to make children aware of a far wider range of models and to try to link the showing of such programs to discussion at school. The importance of the viewing situation discussed earlier suggests that shared viewing and subsequent classroom discussions may well have an effect. Specifically, educational programs on selected aspects of sexual development do make an impact on the development of attitudes. Where they do not, the failure may have to do with content or presentation, that is, with the program itself or with the training and readiness of the teacher to build on the information shown.

We are not talking here about an esoteric matter, but about the fact that children learn from the mass media and that they tend to identify with screen characters of their own sex. In the case of educational television programs, there is need for a good deal of experimentation with different formats for conveying information about sexuality. Such formats need to vary not only with the age of the child or the particular topic—whether, for example, it concerns sex roles or information about bodily functions—but also with the attitude and skill of the teacher, and the atmosphere in the classroom. Relevant factors include the relationship of pupil to teacher, the ease with which the teacher answers (often personal) questions, and talks about sex, sexual feelings, love, and sex roles. Where there is teacher embarrassment, the programs should perhaps stand by themselves. Where the teacher requires only an occasion or a setting to start the discussion, small vignettes that pose a question might be the better solution. It would therefore be very unfortunate if a set of educational programs were provided with the expectation that most teachers would be using the same program. It would also be extremely helpful if the teachers, in previewing the program, could be informed about the range of questions that pretesting produced from school children.

Because stereotyping is excessive, we have devoted a good deal of attention to providing the necessary evidence and to suggesting remedies. It would be absurd to consider that that is all there is to the question of sexuality. In addition to the need for programs that show men and women in different contexts, we need programs that convey sexual knowledge, not in a narrow sense, but more broadly—programs in which sexual mores, the link between sex and affection, and the raising of children are all touched upon. Such programs, providing their effectiveness is carefully assessed, could also be used in schools. They should be used only by teachers who feel comfortable about discussing such topics.

What is needed is not to preach, but to bring information to the child so as to broaden his or her perspective. We suggest that children—like television producers, writers, and decisionmakers—need to examine critically what is offered and to be made aware of their own stereotyped thinking. As is often the case, demand for alternatives, whether to violence or sexual stereotyping, will in the long run benefit the industry itself by offering new challenges to writers and a broader field from which to draw material for their stories.

All too often the television industry has been described as immensely powerful and the viewers as helpless recipients of the material that is doled out. On the contrary, there is hardly an industry where the viewer holds such sway, but his or her appetite for critical judgment, and for making positive suggestions, needs to be aroused. Just as schools teach a critical attitude to books, so they should teach a critical attitude to television. We began this chapter with a quotation from Fearing about the cinema. We should like to end with a quotation written two hundred years earlier (1747) by Samuel Johnson on the occasion of the opening of Drury Lane Theater, London.

> The stage but echoes back the public voice.
> The drama's laws the drama's patrons give
> For we that live to please, must please to live.[62]

NOTES TO CHAPTER 4

1. F. Fearing, "Influence of the Movies on Attitudes and Behaviors," *Annals of the American Academy of Political and Social Science*, (1947): pp. 70–79.
2. S. Schachter, *Emotion, Obesity, and Crime* (New York: Academic Press, 1972).
3. H.H. Kelley, "Causal Schemata and the Attribution Process," in *Attribution: Perceiving the Causes of Behavior*, eds. E.E. Jones et al. (Morristown, N.J.: General Learning Press, 1971), pp. 173–181.
4. L. Kohlberg and E. Zigler, "Physiological Development, Cognitive Development and Socialization Antecedents of Children's Sex-Role Attitudes," in *Sex Differences: Cultural and Developmental Dimensions*, eds. P.C. Lee and R. Sussman Steward (New York: Urizen Books, 1976), pp. 435–443. Our italics.
5. J. Piaget, *The Origin of Intelligence in the Child* (London: Routledge and Kegan Paul, 1953); J.S. Bruner, *Processes of Cognitive Growth: Infancy* (Worcester, Mass.: Clark University Press, 1968).
6. A. Bandura and R.H. Walters, "Theories of Identification and Exposure to Multiple Models," in *Sex Differences: Cultural and Developmental Dimensions*, pp. 423–433.
7. "Children and Television: A National Survey Among 7–17 Year Olds" (Cambridge, England: Carrick James Market Research, March, 1978).
8. T. Furu and J. Lyle, "The Function of Television for Children and Adolescents," *Studies of Broadcasting* 8 (1972): 107–110.
9. G. Comstock, "Television's Four Highly Attracted Audiences," *New York University Education Quarterly*, 10, 2 (1978): 23–28.
10. H.T. Himmelweit, "Education and Broadcasting," *Educational Broadcasting Review* 5 (1971): 45–53.

11. H.T. Himmelweit, A.N. Oppenheim, and P. Vince, *Television and the Child* (Oxford: Oxford University Press, 1958).

12. Schachter, *Emotion.*

13. Bandura and Walters, "Theories of Identification."

14. Comstock, "Television's Audiences."

15. B.S. Greenberg, S. Bradley, and B. Dervin, *Use of the Mass Media by the Urban Poor* (New York: Praeger, 1970).

16. G. Gerber, "Violence in Television Drama: Trends and Symbolic Functions," in *Media Content and Control*, Television and Social Behavior, vol. 1 (Washington, D.C.: Government Printing Office, 1972), pp. 28–187.

17. A. Beuf, "Doctor, Lawyer, Household Drudge!" *Journal of Communication* 24 (1974): 142–145.

18. E.E. Maccoby, W.C. Wilson, and R.V. Burton, "Differential Movie-Viewing: Behavior of Male and Female Viewers," *Journal of Personality* 26 (1958): 259–267.

19. Bandura and Walters, "Theories of Identification."

20. P. Mussen and L. Distler, "Childrearing Antecedents of Masculine Identification in Kindergarten Boys," *Child Development* 31 (1960): 89–100.

21. A. Bandura, D. Ross, and S.A. Ross, "A Comparative Test of Status Envy, Social Power, and Secondary Reinforcement Theories of Identificatory Learning," *Journal of Abnormal Social Psychology* 67 (1963): 527–534.

22. W. Schramm, J. Lyle, and E.B. Parker, *Television in the Lives of Children* (Stanford: Stanford University Press, 1961); G.A. Hale, L.K. Miller, and H.W. Stevenson, "Incidental Learning of Film Content: A Developmental Study," *Child Development* 39 (1968): 60–77; Beuf, "Doctor, Lawyer."

23. Hale, Miller, and Stevenson, "Incidental Learning."

24. Maccoby, Wilson, and Burton, "Differential Movie-Viewing."

25. Beuf, "Doctor, Lawyer."

26. M.M. Miller and B. Reeves, "Dramatic TV Content and Children's Sex-Role Stereotypes," *Journal of Broadcasting* 20 (1976): 35–50.

27. Himmelweit, Oppenheim, and Vince, *Television and the Child*; Schramm, Lyle, and Parker, *Television in the Lives of Children*; Pye, "Report on Patterns of Television Viewing."

28. Greenberg, Bradley, and Dervin, *Use of the Mass Media.*

29. B.J. Calder, T.S. Robertson, and J.R. Rossiter, "Children's Consumer Information Processing," *Communication Research* 2 (1975): 307–316.

30. Himmelweit, Oppenheim, and Vince, *Television and the Child.*

31. Beuf, "Doctor, Lawyer."

32. Himmelweit, Oppenheim, and Vince, *Television and the Child.*

33. D. Roberts, C. Herold, M. Hornby, S. King, D. Steine, S. Whitely, and L. Silverman, *Earth a Big Blue Marble: A Report on the Impact of Children's Television Series on Children's Opinions*, (Stanford, CA.: Communications Research, Stanford University, unpublished manuscript, 1974).

34. L.J. Busby, "Sex Role Research on the Mass Media," *Journal of Communication* 25 (1975): 107–127.

35. C. Cathey-Calvert, "Sexism on Sesame Street: Outdated Concepts in a Progressive Program, (Pittsburgh, PA.: *KNOW*, Inc., 1976.)

36. M.L. Long and R. Simon, "The Roles and Statuses of Women on Children and Family TV Programs," *Journalism Quarterly* 51 (1974): 107–110.

37. L.J. Busby, "Sex Roles as Presented in Commercial Network TV Programs Directed toward Children: Rationale and Analysis," (Ph.D. diss., University of Michigan, 1974).

38. Ibid.

39. Ibid.

40. Ibid.

41. N.I. Katzman, "Television Soap Operas: What's Been Going on Anyway?" *Public Opinion Quarterly* 36 (1972): 200–212.

42. S. Franzblau, J.N. Sprafkin, and E.A. Rubinstein, *A Content Analysis of Physical Intimacy on Television* (Stony Brook, N.Y.: Brookdale International Institute, 1976), p. 14.

43. J.F. Seggar, "Imagery of Women in Television Drama, 1974," *Journal of Broadcasting* 19 (1975): 273–282; "Women's Imagery on TV: Feminist, Fair Maiden, or Maid?" *Journal of Broadcasting* 19 (1975): 289–294; and "Imagery as Reflected Through TV's Cracked Mirror," *Journal of Broadcasting* 19 (1975): 297–299.

44. J.R. Dominick and G.E. Raunch, "The Image of Women in Network TV Commercials," *Journal of Broadcasting* 16 (1972): 259–265; A.E. Courtney and T.W. Whipple, "Women in TV Commercials," *Journal of Communication* 24 (1974): 110–118.

45. E. Wartella and J.S. Ettema, "Cognitive Developmental Study of Children's Attention to Television Commercials," *Communication Research* 1 (1974): 69–88.

46. S. Kline, "Structures and Characteristics of Television News Broadcasting: Their Effects on Opinion Change," (Ph.D. thesis, University of London, 1977).

47. A. Bandura, *Principles of Behavior Modification* (New York: Holt, Rhinehart, and Winston, 1969).

48. S. Ward and D.B. Wackman, "Children's Information Processing of Television Advertising," in *New Models for Mass Communication Research*, ed. P. Clark (Beverly Hills, Calif.: Sage Publications, 1973), pp. 119–146.

49. Although television commercials and television programs have been the most frequent subject of content analysis, some analyses have also been done on children's books. L.J. Weitzman conducted a content analysis of the books that were awarded the Caldecott Prize as well as other children's books. They found that women are under-represented, thus giving the impression that girls are not important since few people bother to write about them. By comparison with boys and men, girls and women have less varied pursuits and are less adventurous and independent. Their activities are generally confined to indoor, and those of the males to outdoor, pursuits. (L.J. Weitzman et al., "Sex-Role Socialization in Picture Books for Preschool Children," *American Journal of Sociology* 77 (1972): 1124–1150.)

Nor do elementary school English readers correct these stereotypes. M.E. Taylor found that in third-grade readers, the central characters were male; five out of eight stories featured a boy with a male adult. The other three depicted lives of men or the adventures of boys. She concludes: "How better to teach little girls their insignificance than by ignoring them altogether?" (M.E. Taylor, "Sex-Role Stereotypes in Children's Readers," *Elementary English* 50 (1973): 1061–1064.)

Of equal importance to studies of books and television would be a study of radio programs—not only of fictional programs but also of phone-in programs where men and women discuss their concerns. In addition, recordings are another source where strong identification occurs. The content of songs often reflects idealized norms of a particular age group. They are today's version of the ballads of the past.

50. R.S. Rogers, "The Effects of Television Sex Education at the Primary School Level," in *Sex Education: Rationale and Reaction* (Cambridge: Cambridge University Press, 1974), pp. 251–264.

51. Grampian Television, "A Report on Living and Growing," in *Sex Education: Rationale and Reaction* (Cambridge: Cambridge University Press, 1974), pp. 227–238.

52. B.S. Greenberg and B. Reeves, "Children and the Perceived Reality of Television," *Journal of Social Issues*, vol. 4 (1976): 86–97.

53. George Comstock, "The Impact of Television on American Institutions and the American Public," *Journal of Communications* 2 (1978): 12–28.

54. P. Rushton, "Social Learning and Cognitive Development: Alternative Approaches to an Understanding of Generosity in Seven to Eleven Year Olds," (Ph.D. thesis, University of London, 1973).

55. Fearing, "Influence of the Movies"; Furu and Lyle, "Function of Television"; G. Salomon, "What Is Learned and How It Is Taught: The Interaction between Media, Message, Task, and Learner," in *Media and Symbols*, ed. D.R. Olson (Chicago: University of Chicago Press, 1974), pp. 383–406.

56. S.H. Chaffee and J.M. McLeod, "Adolescent Television Use in the Family Context," *Television and Adolescent Aggressiveness*, Television and Social Behavior, vol. 3, (Washington, D.C.: Government Printing Office, 1972).

57. J.L. Singer and D.G. Singer, "Fostering Creativity in Children: Can TV Stimulate Imaginative Play?" *Journal of Communication* 26 (1976): 74–80.

58. Salomon, "What Is Learned."

59. Joanne Wawrykow, "An Experimental Study of Vicarious Conditioning Using Film," (Ph.D. thesis, University of London, 1971).

60. Ibid.

61. Kline, "Structures and Characteristics."

62. Samuel Johnson, "The Vanity of Human Wishes," 1749, in *The Oxford Dictionary of Quotations*, 3rd ed. (Oxford: Oxford University Press, 1979), p. 282

Chapter 5

Sexual Learning in the Elementary School

*Michael Carrera**

I. INTRODUCTION

This chapter will examine the elementary school as a major mediating force in children's sexual learning. The school will be studied from the following perspectives: (1) the form, function, and content of school programs called sex education, family-life education, or human sexuality; (2) the form and content of traditional elementary school curriculum activities and the educational materials used to facilitate this learning; and (3) the total ecology of the school setting and its role in communicating sexual learning to children—with special emphasis on the incidental, adventitious, and informal learning about sexuality that occurs in elementary schools.

It is important to underscore at the outset that the school cannot be seen as the only source of sexual education for the child. The family (defined as at least one adult and one child) has been, and continues to be, the primary influence on the child. One cannot discuss the general sexual development of the child without appreciating the profound and compelling value of parents and other family members within a home. While the influence of the family is not absolute, the early identification with parents, first impressions of femaleness and maleness, the value placed on people, and the family's prevailing religious and cultural orientations create an environment that leads to

*In the forefront of public education about human sexuality, Michael Carrera is Chairperson of the Board of Directors of the Sex Information and Education Council of the United States (SIECUS). Dr. Carrera has published widely in the field of sexuality, and is a former president of the American Association of Sex Educators and Therapists.

the development of specific potentialities within the child. These family forces, in my view, produce a foundation that the child brings to the school setting; although the foundation is not unalterably fixed, it must be given serious consideration when evaluating relative influences on the formation of the child's sexual self.

It is also important to recognize, however, the enormous impact of sociocultural forces and environmental forces other than the family on the sexual learning of the young person. The mass media, peer groups, and the culture itself—discussed in other chapters in this volume—are among the most influential of these forces. Additionally, the genetic, instinctual, and biological endowment of the child must also be considered, as must an appreciation for the child's idiosyncratic biological and emotional timetable.

The material presented herein must also be viewed within the bounds of my definition of sexuality. In the past, as has been noted by Elizabeth Roberts in Chapter 1, many regarded sexuality as something entirely apart from daily life. The word itself typically brought to mind some kind of transitory, episodic, and essentially genital experience. Today, otherwise enlightened individuals still discuss sexuality in terms of conception control, unintended pregnancy, adolescent promiscuity, venereal disease, and other behavioral manifestations that support and perpetuate a constraining understanding of sexuality. In short, our society has performed a "sexectomy," that is, sexuality is not seen in relation to the rest of living. In this chapter I use sexuality as a composite term referring to the totality of being a person. Sexuality suggests our human character—not solely our genital nature—and the full meaning of being a man or woman. Sexuality is, therefore, a function of the complete personality and is concerned with the biological, psychological, sociological, spiritual, and cultural variables of life that affect personality development and interpersonal relations. This reconceptualized definition of sexuality is also in keeping with the important World Health Organization definition of sexual health: "Sexual health is the integration of the somatic, emotional, intellectual and social aspects of sexual being, in ways that are positively enriching and that enhance personality, communication and love."[1]

Another concept that requires defining from the outset is that of education. This chapter describes education as a human relations process based on the presentation, exchange, reconstruction, and reorganization of information, feelings, and behavior. The education process is rooted in learning theory and is measurable; its goal is to motivate and sustain the development of social and personal attitudes and decisionmaking behavior that foster self-esteem and respect for the rights and concerns of others.

The process of education is a continuing system of learning, reinforcement, and internalization. Education is a broader and deeper concept than that of cognitive learning alone because education also has an impact on the affective or psychoemotional domain. Education involves the translation of knowledge into healthful practices, the ability to change when personally appropriate, the understanding of alternatives and consequences, and the integration of these processes within a person's life.

II. SEX EDUCATION IN THE UNITED STATES: A HISTORICAL PERSPECTIVE

Most people see sex education programs as a rather current phenomenon that first emerged publicly in the mid-1960s and has increased in popularity and controversy until the present time. On the contrary, however, sex education as an educational concern can be traced back to the late 1880s, when groups such as the YMCA and the Child Study Association sponsored lectures and panels dealing with sex-related topics. In 1892, at its annual meeting, the National Education Association (NEA) discussed the place of sex education in the curriculum. At about the same time, the newly organized National Congress of Parents and Teachers devoted much of its time to discussing the method of implementing sex education in the schools. In 1905, with the organization of the American Society for Sanitary and Moral Prophylaxis, the schools were finally seen as a way to deal with this issue. Although the primary focus of this organization was on the eradication of venereal disease, it was the recognition that this goal could be accomplished through public education that ultimately created the thrust for the development of broader sex education programs.

In the 1880s and through most of the 1890s, sex hygiene, morals education, and other terms used to describe this important initiative were focused primarily on reproductive biology, birth, and the control of venereal disease. For example, the NEA passed a resolution in 1914 urging teacher training institutions to provide training in sex hygiene. In that connection, throughout the 1920s and 1930s, the United States Public Health Service was active in conducting work conferences for teachers on sex hygiene issues. In the 1940s, the United States Public Health Service, in cooperation with the Institute for the Control of Syphilis of the University of Pennsylvania, sponsored a workshop on health and human relations that had an essentially genital focus.

Real pioneering work in sex education did not occur until the early 1950s. Through the leadership of Dr. Lester Kirkendall (who did early work in Oklahoma in 1934) and Dr. Mary Calderone (founder in the mid-1960s of the Sex Information and Education Council of the United States (SIECUS)), a fuller concept of sex education began to receive widespread attention. Kirkendall and Calderone—whose lead was followed by such groups as the American Association of Sex Educators, Counselors, and Therapists (AASECT) and the National Council on Family Relations (NCFR)—provided the necessary guidance for reflection and examination regarding new and important directions for school programs in sex education.[2]

III. FACTORS IN THE ELEMENTARY SCHOOL LEARNING EXPERIENCE

> Every person embodies an adventure of existence and the art of life consists in the guidance of this adventure, an adventure in which men and women must participate equally. The prime business of democracy, the great democratic task of men and women, is not the making of things, not even the making of money, but the making of human beings. . . .[3]

This compelling statement conceptualizes the role that the school ought to play in our society, and suggests the connection between the holistic definition of sexuality and the requirements of the educational processes.

Let us now examine the quality of life that permeates the learning environment in the elementary school. Each dynamic element in this environment is distinct, and yet also inextricably interwoven with every other element. Taken together, these elements constitute a seamless fabric of the total learning process in the elementary school.

In the detailed descriptions that follow, some statements regarding the negative effects of the elementary school on sexual learning may not reflect every reader's experience. However, the conditions presented continue to exist, in varying degrees, in elementary schools throughout the United States and elsewhere. Therefore, it is necessary to raise our level of awareness about the nature and results of such life-distorting experiences, as well as to encourage those practices that have a more positive effect.

The Curriculum Called Family Life Education, Sex Education, or Human Sexuality

Imagine being in the body of an elementary school student and seeing the school day through her or his eyes. Consider what it is like for the student to take part in the school's formal sex instruction program. Characteristically, just what is experienced? Close examination of hundreds of such programs around the United States demonstrates that formal education in sexuality is delivered in one of several ways.

1. The Integrated or Holistic Method. The most appropriate and purposeful method sees sexual learning as an integrated part of the entire curriculum. Since the elementary school child usually has class with a single teacher throughout the school day (except for special activities such as physical education or crafts), this approach is not only reasonable but affords a natural opportunity to examine and learn about sexuality in the context of everyday classroom activities and subjects. This approach deliberately and systematically treats sexual issues and topics simply as they emerge from the variety of subjects being taught. People who use this approach conceptualize sex education in the school—like the student's sexuality—as an integral part of the entire learning experience.

As early as 1967, the integrated approach to sex education was being used in many private schools throughout the United States.[4] Through the leadership of the National Association of Independent Schools (NAIS), private schools have been influenced to include sexual learning areas in their various curricula offerings. Some of the schools responded by considering the topic of sexuality as it naturally emerged from the traditional subject areas.[5] The results of these approaches have been positive, though they have been reported in anecdotal ways. In a recent publication, the NAIS reports continuing success with their entire sex education effort, including the integrated or holistic approach in the elementary schools.[6]

Central to the success of this very sound approach is an understanding of the comprehensiveness of sexuality and its implicit presence in every aspect of learning. Whether it looms as more or less important than other concepts being taught depends, of course, on the teacher. However, recognition of the pervasiveness of sexuality is pivotal to the success of such an integrated program. For example, within the traditional study areas of science, social studies, language studies, mathematics, health and physical education, art, drama, and music, it is possible (if the teacher is prepared and if the ethos of the school is appropriate) for the creative teacher to help develop a climate that fosters learning about the sexual self. The study of the usual subjects in elementary school can naturally include topics of great relevance to sexual learning for young people. Such topics might include: human life, animal life, plant life, privacy, safety, gender and social role development, personality and emotional development, social issues, values and decisionmaking, intimacy, relationships with peers and adults, and physical growth and development.

In such a program a child experiences—at developmentally appropriate times throughout his or her elementary school years—information, concepts, attitude development, and behavior opportunities consistent with the perception that sexuality is implied in all aspects of learning and life.

The treatment of the topics by the teacher, and the learning and activities experienced by the children should emphasize concepts concerning equality of the sexes, equality of aspiration for the sexes, the entitlement to the expression of a full range of emotions for both girls and boys, and the value and normality of relationships with members of the same and the other sex.

In this context, I must point out that a critical part of each of these areas is also the development and practices of an appropriate, humane, and sensitive vocabulary relating to the new knowledge required. It helps little if new information is fixed in sexist and insensitive language.

A major challenge, of course, connected with the success of this important initiative in school sex education is to have teachers available who are sensitive to the holistic definition of sexuality, and who will use the natural opportunities in the curriculum to address the issues as they arise. Most current preservice teacher preparation programs do not provide leadership in this direction.[7] As a result, inservice and continuing education programs are trying to elaborate this important concept, but thus far they have remained fragmented and have achieved little success or distinction.

2. The Nonschool-Based Specialist Approach. In a superficial attempt to legitimate their sex education programs, many schools use nonschool personnel (physicians, health department personnel, or family planning personnel) to deliver a short program on the topic. There are several major disadvantages to this approach. It clearly separates sexuality from the rest of the curriculum, which communicates to the student that this topic is of a special nature and requires a different treatment from other subjects in the school. In addition, students may be unwilling to share thoughts or feelings about sexuality with their teachers if the message is that only special people can discuss the topic. Furthermore, it is unwise for children to associate sexuality with medical and clinical personnel. Children usually see doctors and other medical personnel when they are ill or distressed. Experiencing them in this context in school does not encourage the development of a disease-free, joyful, healthy concept of sexuality. This particular limitation, interestingly enough, has its homologue in medical schools today where sexuality is presented as part of gynecology, urology, and psychiatry but not usually in other clinical disciplines such as medicine, surgery, and family practice.

A further shortcoming of the specialist approach is the need to plan programs around the limited time the specialists are available, rather than around the developmental needs of the children. Finally, the nonschool specialist rarely knows students well enough to develop learning material that is especially suited to particular groups or individuals.

3. Special Sex Education Units with Already Existing Courses. A common approach to sex education in the elementary school setting is to develop a unit of learning about sexuality for inclusion in an appropriate existing course. Science, for example, is usually seen as a natural context for sex education. Many schools, therefore, deal with plant life, animal life, and human life issues in science classes and present information on sexuality at that time. Typically, each year, the student receives a week or several weeks of sex education within a single subject area (science, hygiene or physical education, home economics) deemed appropriate to the material. After completion of that unit of learning, the sexual interests and concerns of the elementary school student must wait until the next academic year for further response. The obvious drawbacks of this approach include, once again, the separation of sex from other topics and issues, and the imposition on sexual learning of a time limit that shows no regard for the developmental and spontaneous needs of young learners. Furthermore, through this approach sexuality becomes associated with a specific subject area—usually one that focuses on the cognitive or technical, rather than on the affective or psychoemotional domain.

4. Separate Course Taught by Special School-Based Teacher. An increasingly common approach taken in elementary schools today is to develop an individual course in sex education and to offer it each year (from kindergarten through sixth grade) at a specific time. This course is usually taught by a special teacher (home economics teacher, physical or health education teacher, school nurse, or another teacher who may have some special training). This approach—though more acceptable than the periodic outside resource or the sex unit within an existing course—advances the notion, once again, that one particular period is "sex time," and that only during this time are issues of sexuality explored and discussed.

The fragmentation caused by this and the preceding views of sexual learning seriously limit the effectiveness of sex education. Unfortunately, administrative unreadiness, the lack of teachers who are prepared and comfortable with the subject, problems with course scheduling, and other elements particular to the ethos of the school have inhibited the wide adoption of an integrated, comprehensive approach to sex education.

Elementary School Curriculum, Teaching Activities, and Materials

1. The General Curriculum and Materials. Learning about sexuality can occur in many subject areas in the elementary school. Unfortunately, much of this sexual learning is either overtly or subtly sexist and gender-role stereotyped.

Sexism is the conscious or unconscious assumption that one gender is inherently superior to the other, with the behavioral implication that one gender dominates the other. Gender-role stereotyping, which is the major outcome of sexism, is the categorization of females and males so that access of each to the development of their interests, potential, and skills depends on preconceived notions of what goes best with each sex.

In elementary schools, young people all too quickly learn who is smart in what subjects, who is strong, who is passive, who is assertive, who can be independent, who is successful, and who should show feelings. In a study conducted by the School Library Association, a content analysis of 154 children's picture books showed that 34 percent of the books excluded women completely. Of the remaining books, when women did appear, their roles were homemaking related in 84 percent of the cases.[8] These picture books portrayed boys building and creating things, using their intelligence and wit, showing initiative and strength. Girls were rarely portrayed as possessing these characteristics or as mastering a skill other than homemaking.

Children's reading books are not the only culprit. Gender stereotypes pervade all materials used in elementary school curricula. For example, my own informal analysis of mathematics books in school libraries reveals that even math problems are presented in social contexts that reinforce gender stereotypes. Boys learn to count and add by pretending to drive cars, fly planes, engineer trains and other such traditionally male activities. Girls learn to add and count by jumping rope, measuring cloth, and weighing and measuring recipe ingredients. Boys' faces commonly portray confidence and mastery, while girls' faces reflect bewilderment (until they've been helped by the boys). Sex-typing in social studies and health education books also indicates what is expected in our society. Stereotyped portrayals of scientists, doctors, nurses, explorers, receptionists, assistants, and many others continue to imprint young people with clear preconceptions of "what is" and "what is expected."

History texts, too, are dominated by male achievements and leadership, while females are depicted primarily as wives, mothers, nurses, and cooks. While these traditionally female roles serve an important function in society, their value has generally been taken for granted, rather than recognized and acknowledged as useful. Furthermore, the narrow viewpoint that would assign only these roles to women ignores the many achievements of women outside the home. Women's contributions to civil rights, medicine, law, culture, and organized labor are not elaborated or portrayed honestly or fully in history books. The American Association of School Administrators addressed this issue in their Executive Handbook Series:

> History books are weighted in favor of the political and military, against the economic, social, and cultural. Because of women's virtual exclusion from direct participation in the first two fields (except for Elizabeth I, Joan of Arc, and a few others), they are all too completely excluded from the history books as well. Typically, the only women mentioned in U.S. history are Betsy Ross, Pocahontas, and Eleanor Roosevelt. The long, bitter struggle for woman suffrage, begun by female abolitionists, is made to appear a genteel pastime rewarded by a chivalrous Congress. In addition, suffrage is presented as the ultimate goal of the women's rights movement. The texts ignore the continuing inequities in areas such as employment, credit, marriage and divorce, and education—as well as the continuing struggle to remove them. Instead, they present a few female writers and artists in isolated sections on cultural background.[9]

The general literature in all elementary school subject areas is quite revealing with regard to the current social and life expectations for each sex. A major study of elementary school books used in all curriculum areas showed the following ratios of material: boy-centered to girl-centered, 5:2; adult male to adult female main characters, 3:1; biographies of men to biographies of women, 6:1; male to female in animal stories, 2:1; male to female in folk or fantasy stories, 4:1.[10]

Ingenuity, creativity, bravery, perseverance, achievement, adventurousness, curiousity, competitiveness, assertiveness, friendship, sportsmanship, generativity, autonomy, and self-respect are all traits regarded as positive by our society. Elementary curriculum materials, however, tend to assume that these characteristics are more typical of one gender than the other. Thus, even general learning units at the elementary level impart values about role expectations that are in keeping with a narrow view of sex and gender roles.

The following conclusions can be drawn from this review of the general curriculum and learning materials used in elementary schools:

1. Females are under-represented as important characters
2. The most serious under-representation is in social studies, science, and math texts
3. Occupational models portrayed in elementary school learning materials strongly favor and represent males
4. Girls are represented as having lower self-esteem, less motivation and lower analytical abilities than boys
5. Girls are represented as being more suggestible and more social than boys.

These conclusions are more than an academic exercise, for they are essentially the conclusions that will ultimately be drawn by the children themselves. The dictates of these stereotypes, ever-present in our society, have a cumulative effect on expectations and aspirations, and it is clear that the longer one lives with a distortion the more difficult and confusing it becomes to modify behavior and gain personal congruence.

For example, the sex segregation and sexual stereotyping that has traditionally existed in physical education and sports activities has had a compelling effect on the body image of both sexes. Girls and boys have been led to believe that their bodies have certain competencies solely as a function of their genders. Therefore, young people present themselves in ways that reflect those messages, and have aspirations and feelings for themselves, and about others that are connected to what they have been taught about the meaning of the female body and the male body. These meanings frequently remain with children throughout their early lives, and can act as potent predictors for how they will feel about their body as adults.

Thus, during the earliest school years of a person's life, the educational process is doing little to nurture uniqueness. Rather, young people are treated as members of a group with some previously assumed average characteristic. This process creates a climate in which young people—and indeed, all people—have very little opportunity to realize their own potential in a way that values individuality and self-fulfillment. These values imply not that there must be equality of outcome—with equal numbers of women and men in each role—but only that there should be the widest variation in possible outcome that is consistent with the range of individual differences among people regardless of sex.

2. **Classroom Teachers.** There is no question that teachers share the conscious and unconscious gender-role ideology that affects society generally. Sustaining society's stereotypes is crippling to both sexes, and teachers who do so in their interaction with children must be quickly and thoroughly discouraged. It is heartening to note that books like *The Sexual and Gender Development of Young Children: The Role of the Educator* are beginning to be used in teacher training institutions.[11] This preservice preparation may begin to make educators more enlightened about the subtle and dangerous effects of sex and gender distortion.

Separating girls and boys in seating patterns, having separate areas for hanging coats, and choosing class helpers by sex for specific chores all call attention to sex distinctions and gender roles.[12] Believing or unwittingly acting on gender-differentiated perceptions of personality—for example, that males are dominant, independent, and assertive; that females are dependent, submissive, emotional, or concerned about their appearance—inevitably leads to educational tracking of the worst kind and seriously limits the expectations and potential of all students.

The physical separation of boys and girls can be interpreted in a variety of ways. At one level, teachers continue this practice because of their belief that a certain degree of mischief and teasing will be avoided. Also, some teachers feel that girls need this kind of protection and insulation from supposedly aggressive boys who might take advantage of and dominate the girls. At another level, however, this kind of separation can make the groups antagonistic towards each other, and inspire attitudes of disdain, fear, and mistrust between them. Another explanation of teacher and school policy that separates the sexes is the fear that boys and girls will become involved in sexual games together. For many teachers, this very terrifying possibility is avoided by turning girls and boys into "natural" rivals, and by accentuating stereotypical behavior differences. This process helps to prevent girls and boys from developing open, healthy friendships with one another.

It appears that teachers must first raise their individual and collective awareness regarding these issues, and then take direct and affirmative action to avoid sexism and gender stereotyping in the classroom. Specifically, teachers need to examine their language and be certain that it is free from the common inappropriate references made in this regard. For example, teachers typically tend to use gender-differentiated terms of encouragement and suggestion to students. Similarly, teachers must encourage both girls and boys to express a full range of emotions. They must also encourage participation in a broad variety of activities both in and out of school. Woodworking, sports, home care, and the like are the domain of both sexes, and teachers need to systematically promote interest in these and other such areas.

Sexual learning in the classroom also takes place through assumptions about family styles made by teachers and reflected in learning materials. Until rather recently, it was not unreasonable for teachers to assume that most students lived with a mother, a father, and one or two siblings. In our changing society, however, a great many children live in single-parent households and family size—that is, number of parents as well as number of children—is shrinking. It is imperative for teachers to be sensitive to the changing definition of the family, with all its implications for the lives of young people.

In addition, classroom teachers are in the best position to promote and support attitudes that allow people of either the same or the other sex to display affection toward one another. Discrimination or stereotyped notions based on age constitute another area in which teachers have a responsibility to guide young people. The sexual learning inherent in these areas is subtle but important.

3. Special Teachers and Special Learning Activities. There are several examples of special teacher and learning activities that convey sexual learning.

In health and sex education classes, sexual learning comes from more than the actual course content. When boys and girls are separated for these classes, they are receiving a message about the secret nature of sex information. This attitude can foster discomfort with their own bodies, and is almost certain to cause misunderstandings and difficulty in the long run in communicating with the other sex. On the other hand, some teachers have made a strong argument for separating young people for some sex education classes, particularly in the older elementary grades. They argue that adolescents have a natural reticence about many sensitive issues at this stage of their lives, and that they feel more comfortable discussing some topics in same-sex groups. This point of view should be seriously considered in planning sex education lessons.

Physical education and sports constitute another important area of sexual learning. Many elementary schools have separate physical education classes for girls and boys, with distinct physical activities for each sex. Girls are often denied access to the traditionally "male" sports, and generally given less encouragement in this whole area. Similarly, boys are not encouraged to participate in such "female" activities as dancing, skipping rope, or using the balance beams. When children are interested or determined enough to pursue activities not seen as suitable for their gender, they are labeled "tomboy" and "sissy." No one knows how many less assertive children would take up these "cross-sex" activities if there were not such a social stigma attached. Furthermore, by creating artificial barriers to experience, schools perpetuate the myths that all boys are athletically inclined and that all girls are not—thereby severely limiting the potential of both sexes.

In order to prevent sex discrimination in sports programs and in education in general, the U.S. government passed landmark legislation—Title IX of the Education Amendments of 1972.[13] The key provision of Title IX mandates that sex discrimination be prohibited in federally funded education programs from the preschool through the university level. Both the letter and the spirit of this law directly affect all subject areas: extracurricular sports and clubs, health and physical education programs, vocational and technical courses, student counseling activities, and other areas in the school program.[14] Since the law went into effect, there has been a rapid increase of activities in which girls and boys at the elementary level are given access to participate in activities and learn together on an equal basis.

Examination of the budgets, equipment, and materials available in elementary schools—especially for physical education and sports—also sheds light on the attitudes and expectations of the young as seen by school personnel. For example, there are generally more male coaches than female coaches, and fewer school-endorsed athletic programs for girls. Similarly, when physical education and sports facilities are separated for boys and girls, boys usually get first claim on the budget and the well-equipped gymnasium, while girls get inadequate facilities and a full supply of jump-ropes. It is not as important that there be an absolute piece-for-piece and space-for-space equivalence as that the opportunities and materials available for learning be person-oriented and not gender-related.

Because options for special activities such as home economics and industrial arts are also usually offered on the basis of gender, a strong message is communicated to young people about the nature of "appropriate" extracurricular activities. These messages are often compounded and reinforced by vocational counselors who employ strong channeling and tracking that is based essentially on gender, and that sometimes leads to unrealistic or unfulfilling career choices in adult life.

These illustrations and many others suggest that many elementary schools are mere vehicles for social sorting and classifying that reflects past and current societal attitudes, values, and expectations about sexuality. Our culture has developed life scripts for females and males, and the elementary school often profoundly sustains and elaborates the limiting effects of this system.

Ecology of the School

1. Administrators, Supervisors, and School Policies and Practices. The general ecology of the school, as distinct from the formal teaching, plays an important part in the sexual learning of the elementary school child. On careful examination, the informal, adventitious learning about sexuality that occurs within the context of general school policies and practices is quite impressive. For example, schools that treat all children with respect and equal attention, rather than separating them by sex for various activities, make a positive contribution toward healthy sexual learning that these young people will carry with them into adulthood.

Some elementary schools in the United States, especially parochial schools, still maintain separate entrances for girls and boys. In addition, some schools have separate play yards and separate physical education and sports facilities, as well as separate seating arrangements in the lunchroom, library, auditorium, and art and music classes. Some of the justifications for these practices have been enumerated above, in the section on classroom teachers. It cannot be overstated, however, that this kind of separation artificially keeps boys and girls away from each other, when for the majority of their subsequent lives the quality of their experience will reflect their ability to function together.

Another area of sexual learning is related to the fact that the American public school system is a major employer. In the elementary schools, the majority of classroom teachers, librarians, nurses, and social workers are female, while the overwhelming majority of superintendents, principals, assistant principals, grade supervisors, and doctors are males. To the elementary school youngster, this simple fact conveys the impression that women are limited to teaching, taking care of books, and mending cuts and bruises. Men—who are generally excluded from these tasks—make decisions that materially influence the school program and routines, as well as serving as disciplinarians. Such a pattern also makes a subtle but clear statement regarding authority, power, leadership, competency, intelligence, and prestige. The cumulative effect of this exposure throughout the school career, combined with other similar experiences in the child's universe, makes a dramatic and penetrating impact on the values and expectations children develop for themselves and for others.

School hiring policies for special activities personnel also serve as sources of sexual learning for elementary school children—particularly if those policies reflect traditional stereotypes. It would be extremely helpful to the learning of young people and to society in general if teachers and staff members were hired to fill school positions irrespective of sex. Librarians, industrial arts instructors, nurses, and counselors, as well as teachers of art, music, drama, and home economics all might be either women or men. The experience of always facing a female in home economics, in the library, and as the nurse, and of always facing a male in shop class perpetuates the association of these activities with total life patterns.

School policies and practices connected to maternity leaves have an immediate and important impact on gender role issues specifically, and sexuality generally. Until just recently, only women were eligible to leave their teaching jobs to care for a newborn child. Men were denied the right to be actively involved in the early nurturing process after the birth of their children. This policy suggested quite strongly the places of females and males in our social system, and the elementary school child was sure to hear this message. In this regard, it is interesting to note that paternity leave was an issue that had to be resolved in the highest courts of the states involved, rather than simply an important expansion of social consciousness about the variety of choices available to all people.

School dress codes for students also influence sexual learning. In some schools, dress codes are applied to girls and not to boys: girls are required to wear dresses in all but the most inclement weather. Although rarer today, similar restrictions on dress have been applied to teachers as well. Girls (and women) are more likely to be described as "pretty" when they are wearing dresses (the very language of "handsome" and "pretty" accentuate artificial gender differences). Dress codes—still very much in effect in many private and nonsecular schools throughout the United States—reinforce the idea that physical and sexual attractiveness is related to clothing, and they are another element in the total learning experience of the child.

Some schools in the United States have policies on where children eat lunch. To expect an elementary school child to return home for lunch each day is to overlook several major social changes in the family form. In an increasing number of families today, for instance, both parents work either full- or part-time. The rising divorce rate also means that many single parents must work during the day to support the family. Providing school facilities where children can eat lunches that they have either purchased or brought from home shows a greater awareness of the changing roles of parents today, and constitutes an important communication about gender roles. However, those parents who choose to be at home during the day might object to such a policy.

2. The School Nonprofessional Staff. People who have worked in school environments at any level, from nursery school through graduate schools, have long recognized the critical value of the services provided by the nonprofessional staff. The effectiveness of many school programs is intimately related to the degree of support and service provided by this staff. Similarly, the nonprofessional staff influences the sexual learning of children in elementary schools in a variety of ways—particularly when they are considered as role models.

Usually the school engineer, custodians, and maintenance personnel are male, while the lunchroom aides, cleaning staff, and secretarial staff are female. Men tend to the gardening, the snow removal, basic repairs, and driving the school buses, while women clean the restrooms, prepare and serve food in the lunchroom, volunteer in the classroom, and serve as school crossing guards.

All of these jobs are crucial to the smooth functioning of an elementary school, and children should be taught to recognize their value. The problem being addressed here is the fact that it is so easy to predict which roles will be filled by men and which by women. Once again, this predictability constitutes a clearcut lesson that is part of the informal learning fabric of the child's elementary school experience. Nor is it necessary to wait until the children choose their adult life patterns to see the effects of this lesson. In the typical elementary school lunchroom, girls volunteer to help clean up, while the boys act as discipline monitors or help with such "manly" activities as removing large trash containers.

3. Parents and Community Groups. The way the elementary school relates to parent and community groups communicates to young people still another aspect of learning about sexuality. A close look at the kinds of activities sponsored by the elementary school for parent and community involvement reveals the expectation that schools have of parents. For example, parent aides during the school day are usually female, and parent conferences held during the daytime are difficult for working parents to attend. At the same time, these activities encourage nonworking women, more than men, to be directly active in the schooling of their children.

Field days, special events, and faculty versus parents sports activities are generally more encouraging to fathers or male involvement than they are to female involvement. Bake sales and other food-oriented special events support traditional stereotypes because women usually are more active in such school-child-parent undertakings.

The membership of Parent-Teacher Associations at the elementary school level is dominated by women, and activities sponsored by PTAs usually attract more women than men.

The solution here is not to eliminate any of these activities, but rather to make a concerted effort to broaden their appeal by holding parent conferences in the evenings, sponsoring mother versus faculty softball games, or having a contest for the best cookies baked by fathers. Such simple efforts could serve as catalyzing agents and become important sources of affirmation and approval for both parents and children.

4. Other Students. With all the other environmental forces affecting the students in such compelling ways, it is no wonder that they relate to each other in a traditional gender-coded fashion. Elementary school boys and girls spontaneously act in patterns that reflect the messages they continually receive about what is appropriate for each gender. From chores in the classroom to sports and special activities, boys and girls relate to each other in such set ways that no alternatives seem to exist. On mixed teams, for example, boys are generally the team captains and the leaders. In addition, the rules of a game may be changed to insure less vigorous activity. During these activities, it is much more usual to see organized female cheerleading; rarely are boys encouraged to participate in this kind of activity in elementary school. In the student government or student councils in elementary schools, the boys generally hold higher offices; girls typically fill the position of secretary. Even at this early age, females learn to be satisfied to support the decisionmaking roles of males.

Slowly but inevitably, as young people move through elementary school, they tend to form same-sex friendships and have fewer other-sex friendships. At the same time, heightened competition arises among boys for the social attention of girls—and conversely, among girls for the attention of boys. This situation interferes with the establishment of healthy friendships with either the same or the other sex. This dynamic is due at least as much to the repeated messages about sex differences as to apparent instinctual drives at that age. In this connection, however, it is difficult to differentiate between what springs from within each boy and girl, and what is coded and imprinted by the treatment young people receive in the elementary schools, and have reinforced in other areas of life.

IV. SUMMARY

In this presentation, I have examined the many ways that sexual learning occurs within the context of the elementary school. The subjective and objective facts of the ongoing, routine, and natural interactions that constitute the elementary school experience make a staggering impact on how young people learn about sexuality generally, and gender and gender-roles specifically. Combined with the traditional values still commonly embraced in many families, in the media, among peers, and in many religious institutions, the narrow sexual learning that affects young people serves to sustain and elaborate already existing gender biases, as well as life expectations that are accordingly based on notions of femininity and masculinity.

In elementary schools, the ethos of the school communicates powerful messages regarding the manner in which girls and boys should interact with members of the same and the other sex. These direct and indirect messages affect the social life pattern of young people while in school, and for many, these patterns form the basis for their future interpersonal relationships, and family role structure and interaction.

The reinforcement of overt and subtle sexual stereotyping in every aspect of the school program interferes with girls' and boys' ability to develop acquaintances, friendships, and intimacy according to their own preferences and needs. Young people are influenced to choose only what has been perceived by adults as right and appropriate for them. It is clear that adult messages about friendships and other interactions (like play) are directed in such a fashion as to avoid "sissy," "tomboy," and other so-called inappropriate behavior. By doing this, adults sustain and perpetuate these false and punitive labels, and do violence to the spontaneity, openness, and varied interests of young people. In addition, these rules of behavior support the notion that feelings and actions are either feminine or masculine, and provide no room or encouragement for individual expression. Admitting vulnerability, seeking help, overtly showing feelings, and other expressions of affect are all evaluated by artificial norms and expectations that are gender-related.

In many instances, this social expectation and self-image tracking system leads to the silent suffocation of individuality. Soon, the elementary schools are replaced by other learning institutions like high schools, universities, and workplaces, all of which continue the familiar patterns learned in early life. Furthermore, it is very common to see women and men unwittingly, yet routinely, repeat in their own families, with their own children, the messages that shaped them when they were young people.

Unless significant changes are made in the total fabric of learning institutions, the self-image of young people, their ability to be nurturing, and to form intimate relationships, and their success as family members and parents will continue to be limited and stifled by insensitive and inhumane sexual learning practices, policies, and attitudes.

This chapter has been largely illustrative, and not exhaustive. However, the overview presented does depict how all those who relate directly and indirectly to the young must be aware of the potent influence that they exert on boys and girls, and on the development of the critical life concepts that influence attitudes, expectations, and choices.

The picture of sexual learning in the elementary schools is not a hopeful one. The effects of the interactions described here should draw attention to the enormous challenge facing those who want to change society's basic orientation about the person. If the elementary school is to be changed (and some change is slowly occurring), each of the other major learning environments must be brought into line with a new conceptualization of what it is to be a woman and what it is to be a man.

Currently, the configuration of learning environments is rather consistently limiting; gender is still a major predictor of what a person can expect to achieve in life. A dominant challenge to change is not only to continually examine how individual social environments and their interactive opportunities affect us, but also to improve understanding of the interrelationships between and meaning of all these environmental forces and sociocultural determinants. A study of the total ecology of life is no small or easy task, but the right of all people to express their sexuality and individuality with understanding, joy, dignity, and positive feelings for self and others will not be validated by anything less.

NOTES TO CHAPTER 5

1. World Health Organization, *Education and Treatment in Human Sexuality: The Training of Health Professionals* (WHO Technical Report Series, no. 572), 1975, p. 6.
2. Michael Carrera, "Preparation of a Sex Education: A Historical Overview," *The Family Coordinator*, April 1971, pp. 91–108.
3. Ashley Montagu, *The Natural Superiority of Women* (New York: Lancer Books, 1953), p. 177.
4. National Association of Independent Schools, *Sex Education Packet V* (Boston: National Association of Independent Schools, 1969), p. 1.
5. Ibid., Parts J and K.
6. Mary Sue Miller and Patricia Schiller, *A Teacher's Roundtable on Sex Education*, (Boston: National Association of Independent Schools, 1977), p. 55.
7. Michael Carrera, "Training the Sex Educator: Guidelines for Teacher Training Institutions," *American Journal of Public Health* 62, no. 2 (1972): 233–243.
8. Margaret Higgs and John Stewig, "Sex Role and Sexual Stereotyping in Children's Picture Books," *School Library Journal*, January 1973, p. 54.
9. *Sex Equality in Educational Materials*, American Association of School Administrators Executive Handbook Series 4 (Washington, D.C., 1975): 4.
10. Judith Stacey, Susan Berhand, and Joan Daniels, eds., *And Jill Came Tumbling After: Sexism in American Education* (New York: Dell Publishing Co., 1974).
11. Evelyn Oremland and Jerome Oremland, eds., *The Sexual and Gender Development of Young Children: The Role of the Educator* (Cambridge, Mass.: Ballinger Publishing Co., 1977), pp. ix, 293–305.
12. National Education Association, *Sex Role Sterotyping in the Schools* (Washington, D.C.: National Education Association, 1973), p. 5.
13. U.S., *Statutes at Large* Education Amendments, sec. 901–907; U.S., *United States Code*, sec. 1681–1686, 1972.
14. Margaret Dunkle and Bernice Sandler, "Sex Discrimination Against Students: Implications of Title IX of the Education Amendments of 1972," *Inequality of Education*, no. 18 (1974): 15–17.

FURTHER READINGS

American School Health Association, *Growth Patterns and Sex Education.* Kent, Ohio: American School Health Association, 1974.

Belotti, Elena Gianni. *What Are Little Girls Made of?* New York: Schocken Books, 1976.

Bernstein, Jean. "The Elementary School: Training Ground for Sex Role Stereotyping," *Personnel and Guidance Journal* 51 (1972): 97–101.

Deiseach, Donal. *Family Life Education in Canadian Schools.* Toronto: Canadian Education Association, 1977.

Fisher, L. and Cheyne, J. *Sex Roles.* Toronto: Ontario Ministry of Education, 1977.

Fraiberg, Selma. *The Magic Years.* New York: Charles Scribner's Sons, 1959.

Harrison, Barbara Grizzuti. *Unlearning the Lie: Sexism in the School.* New York: William Morrow and Co., 1974.

Lane, Mary. *Learning About Life.* London: Evens Brothers, 1973.

Maccoby, Eleanor and Jacklin, C.N. *Psychology of Sex Differences.* New York: Basic Books, 1974.

Pottker, Janice and Fishel, Andrew, eds. *Sex Bias in the Schools.* Cranbury, N.J.: Associated University Press, 1977.

Sprung, Barbara. *Non-Sexist Education for Young Children.* New York: Citation Press, 1975.

 *Chapter 6*

Peer Communication and Sexuality

Everett M. Rogers and
*Sharon L. Strover**

INTRODUCTION

As adults, we are quite accustomed to a world peopled largely by our peers. We spend time talking with our friends, and rely on their judgments and advice. We interact with peers socially, and we usually spend our work hours near or with peers. As our social equals—that is, as people like ourselves in status, age, and aspirations—peers reflect our opinions and beliefs. Young children are self-oriented and rely on peers for playmates but for little else. As children grow older, more time is spent in the company of peers, and the friendship group assumes a number of new functions. It can bestow status and approval on the individual; it enforces or creates social mores (such as what to wear, or how to act on a date); it determines attitudes and encourages certain behavior. Many such influences involve aspects of sexuality.

The connotations of sexuality are broad, and—depending on one's perspective—nearly every activity can carry sexual meaning. So, too, an individual's "sexual learning" occurs in many settings—in school, among family, and among friends. While no studies have specifically examined peer group influences on sexuality, it is possible to look at

*Everett Rogers is Director of the International Program at the Institute for Communication Research, Stanford University, and Janet M. Peck, Professor of International Communications. He has written extensively on communication and social change and is coauthor of the recently published book, *Communication in Organizations.*

Sharon Strover is a Research Assistant at the Institute for Communication Research, Stanford University.

the activities and communication styles unique to youthful peer groups in which sexuality is learned and experienced.

The purpose of this chapter is to synthesize what is known and to suggest what should be studied about the peer environment in regard to learning about sexuality. First, the importance of peer groups across developmental stages is examined. For example, in the early years the play group fosters development of role-taking and role-defining skills—some of which are relevant to sex-role orientation and concepts of what a male does and what a female does. Later, teenage activities such as "cruising," dating, and team sports either provide a scenario for exchanges about sexuality or themselves constitute sexual expression. This chapter concentrates on adolescence because it is a period when much social learning about sexuality takes place among peers.

The second section of this chapter addresses the contribution of peer groups to sexual learning, with special attention to sex behavior. A great deal of research has investigated teenage sex, contraception, and pregnancy. This research has found that peer groups are important information and attitude sources for individuals on matters of sex. Neither parents nor schools provide needed advice or knowledge; moreover, sex seems to be a taboo topic in all but the peer group. Hence, peer networks comprise the sex information channel for adolescents.

The final section delves into society's response to the "problem" of adolescent sex—an area under scrutiny because of rising pregnancy and abortion rates for thirteen- to seventeen-year-olds. Three types of social service programs dealing with teenagers or teenage sexuality are discussed in light of what we know about adolescent peer groups and individuals. As we understand the social milieu in which adolescents find themselves, we may be able to bring their concerns about sex and sexuality within range of discussion and understanding.

PEER COMMUNICATION ABOUT SEXUALITY AT VARIOUS STAGES OF DEVELOPMENT

Early Childhood (Birth to Three Years of Age)

The young child engages in solitary play before two years of age, and from three to six begins to play with both same- and cross-sex peers. Until age three, children do not actually experience peer *relationships*. Piaget concluded that until children are about eight years of age, they cannot communicate with each other clearly because younger children do not empathize with their listeners.[1] Other researchers have discussed the important developmental step the child takes in recognizing the self as object and in taking the role of the other. In early childhood this simply does not occur. Nonetheless the activity of play can provide a medium for developing social rules and values that the child practices, elaborates, and incorporates into nascent attitudinal and behavioral patterns. As we illustrate below, the style and content of much play convey sexual messages.

Middle and Late Childhood (Three to Seven, Eight to Twelve Years of Age)

Peer group socializing influence begins in middle childhood and continues through adulthood. By age six, and into early adolescence, unisexual association is the norm, and strongly differentiated groups for males and females form.[2] In addition to playing with each other, children of these ages also attend school together, talk with each other, form close friendships, and generally begin to augment their learning from the home setting with learning from peers. Segregation by gender grows stronger in middle and late childhood, as older children typically choose sex-typed toys, activities, and games.

Sex-differentiated play styles in this age range teach many skills. For example, Janet Lever investigated sex differences in play with respect to (1) the complexity of games, and (2) concomitant opportunities for learning social roles through play and games.[3] Among a group of fifth graders, she examined levels of cooperation or competition in games, size of play groups, whether or not teams were needed, types of roles required, rules used in playing, and player interdependence. A significant finding was that play activities of boys are more complex than those of girls, "resulting in sex differences in the development of social skills potentially useful in childhood and later life."[4] In the same study, Lever also found that boys' play involves more role specialization, more player interdependence, more goal direction and rules, larger groups, and more team division than does girls' play. Consequences of such gender differences may be that girls and boys develop different social skills. Boys' play fosters strategic thinking, skill development, competition, face-to-face confrontation, and the ability to work for team goals; girls' play fosters an indirect style of competition, inexperience with leadership roles and strategizing, and a preference for small groups. In fact, girls more often prefer to be around one good friend, while boys prefer to be in groups. According to Lever, "boys develop the ability to take the role of the generalized other while girls develop empathy skills to take the role of the particular other."[5] The implication is that boys' game experience prepares them for interaction in occupations that share structural features with their games—high levels of organization, teamwork, and competition.

Stanley Aronowitz makes a case for the learning of sex roles in playing games.[6] Examining the role expectations and assumptions in children's games, he notes that the game "The Farmer in the Dell" enables a child to learn typical male and female roles: the man as worker and the female as mother. Games such as "Doctor" and "House" or "Cowboys and Indians" provide a stereotyped scenario in which children reproduce socially expected roles, including gender-specific roles. The boy is the doctor, the girl is a nurse; the boy is the husband and father; the girl is the wife and mother; she does the shopping, while he goes to work. In integrating the adult world into play, children allocate the powerful roles to older or male children. On the other hand, "girls always play mothers and children, nurses and patients, and Indians within heterosexual groups."[7]

Games of older children reinforce appropriate role standards. In late childhood, while girls stay in smaller group games, boys play more team games—such as basketball, football, and baseball—that require competition (with another team) as well as cooperation (within their own team). The emphasis of male sports on winning prepares boys for the adult male world where the name of the game is to get ahead—by competing. Girls too will eventually compete—for partners. The contribution of girls' games to their adult role, though, may lie in the inculcation of cooperativeness and in learning to give attention to only a few people or to one person.

Girls also take the role of spectator to games earlier than do boys. Throughout puberty and beyond, as boys are playing team games, girls move into cheerleader and nonparticipant roles. Such sideline activities are not quite passive, but are certainly accorded far less prestige than actually playing the game.

Iona and Peter Opie document among young children extensive styles and topics of play behavior and linguistic expressions that reflect an awareness of the other sex.[8] For instance, a film star of the opposite sex is a "smasher." "Sweethearts" are special friends of the opposite sex, and a number of endearment terms are used by children of both sexes to denote this special relationship. Games and language specific to children are a form of cultural transmission that is outside of—although sanctioned by—the adult world. While parental or adult control over children's information and reinforcement is dominant, children's peers carry on a form of socialization.

Interest in the other sex sometimes takes the form of group activities. Young boys frighten girls after dark or on their way home from school; groups of girls tease boys. In later childhood, boys and girls develop "crushes" that they talk over with each other. Girls have slumber parties where they talk about boys, parents, and school. Certainly boys' teasing of girls (at the stage of development where anything having to do with girls is anathema) indicates that boys are aware of girls, and of sexual differences. The cult of "macho" behavior probably has its roots in peer groups for older boys that offer an arena of spectators to whom the male can prove himself (without negative adult or female sanctions against unacceptable aggressive behavior). Peer groups clearly operate as sounding boards for emerging conceptions and actualizations of maleness and femaleness. How the individual uses the peer group, however, and to what extent the peer group influences the individual are unknown.

Nonverbal communication among same-sex and cross-sex peers reveal patterns similar to those of adult interaction. For example, Jacob Lomranz and his coworkers studied children aged three, five, and seven and found that at all ages less spatial distance was kept among girls than among boys.[9] The use of space may be an analog for power: the powerful occupy more space and control it against subordinates. According to Nancy Henley, "In general, females in our society have control over less territory, and less desirable territory, than do males."[10] Female territory is also frequently "invaded" by others. A mother's territory may be invaded both by her children and by her husband, who assert control over it at will. Since even a young girl's space is less extensive than her male counterpart's, Lomranz's study might indicate that such sex-differentiated spatial relations are learned at a young age. Reinforcing this idea is the observation that young girls are more frequently touched than are boys by their mothers—with touch constituting variation on personal space.[11] Such different treatment may be training females to get accustomed to the idea of having less control over space.

Once unisexual play groups form in middle childhood, do we find differences between boys' and girls' peer groups? Are the values of boys' groups different from those of girls'? The answer is not yet clear. When peer orientation is tested against individual or adult values, boys are more likely to conform to peers' than are girls.[12] Moreover, girls may feel more comfortable than boys taking on a wider variety of traditionally masculine and feminine behavior. Interpreting this is difficult. Do girls perceive masculine roles to be as desirable as feminine roles? Are girls more androgynous than boys, or does the sex-typing process occur earlier or more strongly for boys? Is conforming to peer group norms rewarded more for boys than for girls?

Two other sources of socialization, the mass media and schools, enter older children's lives (see Chapters 4 and 5 above). Children spend considerable amounts of time with mass media. Research evidence shows that television viewing starts around age two and increases sharply during middle and late childhood. TV viewing decreases during adolescent years as movies and radio and more cross-sex peer interaction replace time previously spent watching TV. Much of the content of television, comic books, and movies present sex-typed models for children to imitate.

Schools, with their tracking systems that reproduce social class systems, determine with whom children interact. Peers are children in the same track, who share the same occupational aspirations and the same values—although sex differentiation in aspirations is evident. The structure of the school usually assiduously avoids dealing with nascent sexuality in school children. However, school texts are notoriously full of gender-specific role models, curricula are geared to generating sex differences (home economics for girls and industrial arts for boys, for example), and teachers may treat the two sexes differently (encouraging boys to be aggressive and girls to be passive).

Adolescence

Peer groups and sexuality both assume greater importance for the individual in adolescence. Childhood socialization lays the foundation of information and attitudes that influence future sexual learning and behavior, but in adolescence the erotic dimensions of sexuality become a conscious, salient part of one's personality. Same-sex peer groups still exist, but teenagers begin to pair off with members of the other sex as well. Additionally, the start of puberty and its biological changes can bring new awareness, drives, and self-consciousness.

Biological Changes

The development of pubic and other body hair, breasts, and larger male genitalia, as well as the onset of menstruation for females signify the presence of erotic capabilities. James Maddock notes that "with the onset of physical maturation, the pressure to conform to sex role behavioral stereotypes become especially intense. There is an intensification of body awareness during adolescence, based upon the fact that the body is a primary 'symbol of self' in which feelings of personal worth, security, and competence are rooted."[13] Many researchers have extrapolated from the fact of body changes that adolescents somehow come to measure their adequacy (as male or female) against prevailing cultural standards of the ideal male and female. Boys and girls may begin to compare their bodies to those of movie and TV stars. Certainly body changes prompt some examination of others to see if such changes are happening to them also—the easiest comparisons being within one's same-sex peer group. "The major reference for sexual conformity is the individual's own sex group. This is perhaps why the young adolescent begins sexual experimentation, both physically and symbolically (verbally) with members of their [sic] *own* sex, and the same sex reference group remains vitally important throughout adolescence."[14] Opportunities for such comparison are frequent. Boys, for example, compare themselves with each other in sports and in locker rooms; girls wear each other's clothes and observe fashion models in magazines and on television.

Social Changes in Adolescence

In addition to biological changes, adolescence is accompanied by social changes—many of which serve to make the peer group a cohesive, primary reference for teenagers.

Adolescents, as a distinct social group, are a twentieth-century phenomenon. G. Stanley Hall first used the word "adolescence" in 1904 to refer to a period of time between childhood and the assumption of adult responsibilities. As a developmental stage, adolescence was a new concept. Previously, an individual had been thought to go directly from childhood into the labor force or marriage—thereby effectively acquiring adult status. Child labor laws, however, gradually raised the age of labor force entrants. Also, with increased mechanization and industrialization, the United States required a labor force with more brain than brawn, so that emphasis was given to lengthened schooling.[15]

Secondary schools really "created" adolescence in the United States. The proportion of children staying in school after age twelve or fourteen grew during this century. By the 1940s, "A distinctive pattern of life began to emerge, around the high school."[16] The pattern was documented by Robert and Helen Lynd in *Middletown*; August Hollingshead's later study of Elmtown observed a peer culture and its informal set of conventions among adolescents.[17] In his examination of the growth of adolescence, Joseph Kett notes that a type of adolescent subculture developed after 1920 that "expressed itself in the construction of a system for evaluating peers that bore little resemblance to the criteria used by middle class adults for evaluating either each other or young people."[18]

The insulation of adolescence from mainstream adult society has increased over the present century, and is growing still. Today adolescents are segregated into special institutions, are confined by special rules, and are deprived of economic wherewithal and contact with people of other ages.

Peers are naturally the crux of the so-called youth culture. Of the three main socializing agents—family, schools, and peers (one might add mass media to this list)—the family is a "minor part of the social environment of many youth beyond early adolescence."[19] As for schools, "the way of life we have institutionalized for our young consists almost entirely of social interaction with others of the same age and formal relationships with authority figures."[20] In other words, for adolescents, the socializing influence of the family and of schools shrinks, and peer group influence expands.

What roles do peer groups play during adolescence? Adolescents use one another as cultural references; they experience a psychic attachment to others of their own age; pressing toward autonomy, they respect those who are independent of adults.[21] The latter may be a particularly post-1960 phenomenon—one prompted or at least facilitated by a strengthened self-definition of "youth" as a subculture. In 1954 D.P. Ausubel et al. signaled many functions of the peer group that may be more important now, some twenty-five years later.[22] They found that peers:

1. Attenuate the importance of status derived from parents, and increase primary status
2. Are a source of derived status
3. Provide new frames of reference to relieve the disorientation and loss of anchorage caused by abandonment of childhood roles
4. Facilitate seeking new sources of value outside the home
5. Are a source of strength in combatting authority
6. Are "the major training institution for adolescents"
7. Provide regularized media, a set of norms for sexual behavior, and occasions for heterosexual contact
8. Reduce frustration and stabilize the individual in this transitional period.

The researchers also wrote that peer groups, in their "nature, structure, norms and purposes are largely conditioned by the characteristics of the particular adult culture or subculture in which [they are] embedded."[23] Such conditioning may be less powerful today.

Peer group functions are important in easing one's transition from dependence on the family to autonomy. Indeed, as adults, we rely on peers probably more than on any other group for continuing feedback on our performance, ideas, attitudes, and self-concept. Peer groups in adolescence provide support to the individual who is breaking away from an identity labeled "child." The group environment shapes where and with whom adolescents spend time, what they do, and how they react to situations and ideas.

The peer group is partly responsible for life-style changes from age thirteen onward. Use of leisure time shifts to accommodate more involvement with people of the same age.

One such leisure time shift occurs in mass media use patterns. Watching far less TV than they did when younger, teenagers use the time instead to cultivate social relationships. Furthermore, teenagers integrated into peer groups use the media differently from those who are less integrated into peer groups. The latter use media for fantasy and are attracted to violent and action themes on TV. Adolescents who are strongly oriented to parents rather than to peers show the highest TV consumption.[24]

Other leisure-time activities also change. In adolescence boys engage in more athletics, begin to date, go to movies, and drive. Girls also date and go to dances, see movies, spend time with friends shopping and listening to records.[25] Movies and music garner considerably more attention in adolescence.

Another life-style change occurs in where peer interaction takes place during adolescence. With the mobility of cars and more independence from parental control, teenagers begin to hang out in such favorite spots as record shops, street corners, neighborhood parks, and recreation sites. Gerald Suttles' study of a slum area in Chicago reported gender-segregated spatial arrangements.[26] Peers of the same sex share a location; there are few cross-sex meeting grounds. Teenage boys take to the streets and frequently are members of gangs, while teenage girls appear to stay fairly close to home. Playgrounds are set aside for male groups or games; girls who invade the boys' playground cast doubt on their character and invite sexual attention.

Although the peer group's influence on values is less discernable, some researchers argue that the peer group operates not so much to undermine parents' relationships with their children as to selectively replace elements of the adult value system with new values.[27] Peers catalyze such substitution, particularly in the area of sexual behavior. Sexuality and sex are frequently difficult topics to discuss openly in the family setting. Their forbidden nature can create a vacuum in the adolescent's value and information system. This vacuum is readily filled with the nearest creditable ideas—ideas often originating from peers.

Thus, the importance of peers—particularly in their influence on notions of sex and sexuality—enlarges during adolescent years. Insofar as the peer group regulates and evaluates one's activities, peer effects on sexuality cannot be overlooked. Peer groups define appropriate places for interaction and guidelines for actions, and they reward conforming behavior. Peer communication figures in transmitting sexually-related information because parents, school teachers, and other adults are often perceived by adolescents as opposed to communication about this topic.[28]

In Table 6–1 we summarize the relative importance of different communication channels by stages in development.[29] Few researchers have specifically dealt with the importance of these channels in conveying information about sexuality to individuals after childhood. However, in the early developmental stages they assume these channels do contain some content about sexuality, even though their explicit or conscious importance to the child may be low.

The next section of this chapter focuses on the importance of peer communication about erotic conduct. This topic has received more research than sexuality as a whole. What we know of adolescent sexual behavior illustrates the process of communication and the impact of peer communication in shaping sexuality.

ADOLESCENT PEER COMMUNICATION ABOUT SEX

Why are parents relatively unimportant in communicating information about sexual intimacy and eroticism to adolescents? Several studies have shown that only half of all parents give explicit instruction about sexuality to their children.[30] Information from those parents who do so is often perceived as inadequate by their children. About 71 percent of the adolescents in a national survey could not talk freely about sex with their parents.[31] Only 18 percent of the thirteen- to nineteen-year-old boys, and 31 percent of the girls, had ever talked with their parents about birth control; 18 percent and 16 percent, respectively, about masturbation; and 24 and 33 percent, respectively, about venereal disease. As might be expected, adolescents tend to disclose information about sex-related matters to their same-sex parent when they do so at all.[32] Nonetheless, the absolute amount of such disclosure is relatively small. While parents in general are not as important as peers for adolescent socialization, parental influences are especially strong on single sons and on socially isolated sons and daughters, who are not well integrated into a peer network.[33]

Table 6–1. Relative Importance of Communication Channels by Stages in Development.

	Stages in Human Development			
Communication Channels	*Early Childhood 0–3 years*	*Middle Childhood 3–8 years*	*Late Childhood 8–13 years*	*Adolescents 13–18 years*
1. Parents	Dominant	Important	Important	Less important
2. School teachers		Important	Important	Important
3. Peers				
(a) Same-sex		Somewhat important	Important	Very important
(b) Cross-sex				Very important
4. Mass Media				
(a) TV		Very important	Important	Less important
(b) Print		Important	Important	Important
(c) Radio and music				Very important
(d) Film				Important

From past research little is known about the possible conflict between parental influences and those of the peer network. Perhaps such conflict may be minimized by the tendency of sons and daughters to choose friends who share the attitudes of their parents.[34] This particular tendency, however, seems to be greater for children than for adolescents, so the question remains.

Peers as Sources of Sex Information

Peers function as information-givers on sex-related matters and offer moral support in first erotic encounters. In a 1953 survey of males aged ten to twenty, male companions were found to be the most important sources of sex information on such subjects as where babies come from, ejaculation, contraception, menstruation, intercourse, venereal diseases, and prostitution.[35] Using sex experience as an indicator of peer effects, Patricia Miller and William Simon found that male adolescents who have lower peer involvement and higher parent involvement also have relatively lower coital experience. Surprisingly, this trend did not hold for females. The researchers postulated a different peer group role for the two sexes. The male peer group approves of erotic sexual activity and cheers it on; the individual male refers to the group for guidance and encouragement. The female, however, tends to rely on her partner to define erotic activity as a legitimate experience. She does not find in her peer group the kind of support for social behavior and expression available to males. In fact, since the female peer group fosters positive valuation of the romantic but not the erotic, "a commitment to the sexual may interfere in the peer group for females."[36]

Inadequate Sex Information

An important research discovery is that many younger, sexually active adolescents have inadequate information about sex and may not be aware that intercourse and becoming pregnant, for example, are related. Knowledge about contraception is also often superficial, incorrect, and prone to rumor and misinformation.[37] Rising numbers of teenage abortions attest to this problem, although certainly information alone cannot prevent pregnancy. Adolescents have a very limited vocabulary of sex terms, do not know the standard (that is, adult) terms, and find it difficult to read about sex and reproduction.[38] Basic expressions, like "sexual intercourse" and "virginity," are not widely understood by adolescents.[39]

These findings suggest (1) that the level of information about the erotic dimensions of sexuality is low among adolescents, and (2) that their terminology differs from adult, "expert" sources. Dependence on peer communication for answers means that the quality of adolescents' information about sex and sexuality spirals into even more inaccuracy because an adolescent's peers are unlikely to be much better informed than he or she is about these matters.

Adolescent Peer Networks

One reason for the relatively low technical competence of adolescent peer communication about sexually related topics lies in the nature of peer networks. Research on adolescent peer communication indicates that most dyadic links are between homophilous pairs.[40] Hence, adolescents with relatively low levels of knowledge talk with highly similar peers. Such horizontal patterns of interpersonal communication amount to sharing limited knowledge, misinformation, and ignorance.

Furthermore, the personal communication networks of most adolescents tend to be interlocking rather than radial.[41] An "interlocking personal network" is one in which an individual interacts with a set of dyadic partners who interact with each other; thus, one's friends are friends of each other. Such closed, ingrown networks are of very limited value in obtaining new information. One's peers only know what one already knows.

In comparison, a "radial personal network" is one in which an individual interacts with a set of dyadic partners who do not interact with each other. Such an open network is ideal for obtaining and sharing information. Unfortunately for the peer education of adolescents about sex, radial personal networks are infrequent at that age.

Penetrating Peer Networks

Notwithstanding the general tendency, peer communication among adolescents is not completely homophilous. Adolescents interact with some "near-peers," older siblings, parents, and other adults. Moreover, they have exposure to the mass media[42] and to various social service agencies. A certain amount of sex information from other sources does, therefore, penetrate the adolescent peer network.

When some members of a peer network attain puberty at different times, perhaps early maturers are accorded prestige by their peers, and if such early maturing individuals possess greater sex-related knowledge (perhaps based in part on their advanced physical development), it may be an important input into the adolescent peer network.

During adolescence, cross-sex peer interaction increases and may often link older males with somewhat younger females—at least in couples who are dating or going steady. Cross-sex interpersonal communication may be one important type of near-peer information transmission about sexual topics. It is known, for instance, that girl friends are "significant others" in their influence on adolescent males' career aspirations and achievements.[43]

Generally, at age seventeen, peer influences (both male and female) are greater influences on educational achievement than are parental influences. Cross-sex peer influences are probably equally strong influences on sex behavior.

Finally, gender differences might be found in the communication structure of peer networks. Girls may have smaller personal networks with more intensive relationships; boys may have larger, more extensive personal networks.[44] If indeed males have more radial personal networks, and are linked to females with (otherwise) interlocking personal networks, such cross-sex dyads could be important links in transmitting sex information.

IMPLICATIONS

The Social Problem of Adolescent Pregnancy

Many social service agency personnel are painfully aware that the services they provide—whether involved with job banks, suicide prevention, sex counseling, or drug education—are little more than Band-aids on wounds requiring major surgery. Social services involved in delivering sex information or counseling are no exception to this statement. Such services often touch only the problematic "tip of the iceberg" of sexuality; their information and devices skirt the sensitive topics and avoid in-depth discussion. Frequently, too, these services are offered in crisis center settings. Young women go to Planned Parenthood clinics because they believe they are already pregnant; individuals learn about venereal disease only when they have it.

By separating them from other aspects of life, many social service agencies have also "medicalized" sex and sexuality, and treated them as mainly biological or medical phenomena. These agencies overlook the strong influences of society and of group and psychological values on sexual behavior. Recent developments may force an end to this narrow perspective. Adolescents are engaging in intercourse and becoming pregnant at younger and younger ages. According to the

Guttmacher Institute, rates of sexual activity by age in 1974–75 are as follows:[45]

Age	*Percent Sexually Active*
13	10
14	17
15	24
16	31
17	35
18	43
19	51

While pregnancy rates for all age categories in the United States are decreasing generally (although the past two years may have shown a slight increase), pregnancy rates for adolescents are decreasing most slowly of any age category. About one-fifth of all U.S. births belong to adolescent mothers; approximately 10 percent of all adolescents become pregnant each year; 60 percent carry their pregnancy to birth. Abortion rates for teenagers rose by 60 percent from 1972 to 1975—the greatest increase in abortion demand for any age group.

Such statistics have led to treating sexual activity among adolescents as a social problem—a perspective with disadvantages as well as advantages. The realization that sexual understanding involves more than services, contraceptive information, and abortions, and that its importance cuts across race and income levels, may lead to new perceptions of the meaning of sexuality and how society educates the young about it.

Intervention Strategies

Recent trends in social service agencies may ameliorate traditional problems in dealing with adolescent sex by coordinating information and services about sex and sexuality with the clients' personal values. For example, in answer to the Carter administration's preference for "alternatives to abortion," nine leading health and family planning organizations in the United States recommended expanding contraceptive programs to reach 1.5 million more adolescents by 1980 *within a milieu central to their daily lives.* They recommended counseling services, and increased contraceptive and sex education within the schools, in community agencies, and in the mass media. Hopefully, such activities will have the additional benefit of demedicalizing sex and sexuality.

Hotline Services

The initial impetus for peer counseling services owes much to the spontaneous genesis of youth hotlines in the 1960s. The two primary features of such services were the promise of anonymity to the client (facilitated by the use of the telephone and first names; motivated by fear of police surveillance) and a young, uncredentialed staff of volunteers who were near-peers to the client. These made hotlines an alternative service to establishment programs.[46] Hotlines were organized to help with drug withdrawal, prevent suicides, establish contact with runaways, and allow open discussion of sexual matters. Complementary "drop-in" centers for individuals who wanted medical services or personal counseling also sprang up in the 1960s. Such services may have been one part of an attempt to create an alternative youth-based community, a priority among adolescents in the late 1960s.[47] Information transmission by such peer services was essential to the fluid nature of these youth communities.

As hotlines began referring clients to standard social services for help that the hotline could not offer, the more established services realized that their youthful clients needed special communication techniques. One of the most effective was the peer-based provision of information and services.

Hotlines and noninstitutional social services still exist in the 1970s, although not in their previous numbers. Today, many other social service agencies (usually government-supported) have picked up the hotline's function, and its emphasis on near-peer communication. Services specifically related to sexuality, however, are still located in clinics, and provide services about sexuality in a medical, problem-oriented setting that is not conducive to treating sexuality as a lifelong, value-centered aspect of living.

Sex Education in Schools

An example of what sexuality education should not be can be found in our schools. Of the twenty-nine states plus the District of Columbia that require public schools to teach health education, only six states and the District of Columbia mandate sex education in schools. In such classes the content of sex education is so strictly monitored that sensitive subjects such as intercourse, venereal disease, masturbation, and homosexuality are usually banned from discussion. In general, sex education classes do not integrate information about the sex act with other aspects of sexuality such as gender-role stereotyping and values. In fact, it is not very surprising that exposure to sex education in schools is not related to premarital sex behavior or unwanted pregnancy.[48] Currently then, education in schools does not seem to have much of an impact on what an adolescent learns about his or her sexuality or sexual behavior.

Nonetheless, we believe that sex education in schools could be very important. Schools' accessibility and credibility are unequaled. Currently lacking are (1) support from communities for in-depth sex education courses and (2) curricula that include the "other-than-plumbing" material. Sex education in schools can solicit parents' input, and give students the opportunity to discuss with peers and a sensitive teacher some of their more intimate questions, as well as such sexual issues as homosexuality and sex roles (see Chapter 5 above).

Peer Counseling Programs

Peer counseling is a technique that promises deeper attention to individuals seeking information about sex. Peer counseling may help demedicalize sexuality services through (1) the homophily of the providers and (2) the nonhierarchical setting.

Peer counseling is based on the kind of information-seeking behavior of adolescents that was examined earlier in this paper. Information about sex is obtained mainly from friends and the media. While school personnel, parents, and other adults are of less importance as information sources during the teen years they are the sources for many values that determine how information about sex is used. A mechanism is needed to integrate these underlying values with specific information. Peer counseling offers promise of doing exactly that.

Most social service agencies that are currently using peer counseling deal only with a rather narrow aspect of sexuality, that is, sex information. In many cases, the agency's interests and service capabilities are not amenable to tackling the broader aspects of sexuality. Where it has been tried, the advantages of peer counseling about contraception and sex have been found to be effective, economical, and acceptable both to parents of adolescents and whatever institution houses the counseling (predominantly schools and clinics).[49]

The first peer counseling program for sex-related services was established as recently as 1969, by Dr. Sadja Goldsmith at Planned Parenthood in San Francisco. This program used near-peers—people who were just beyond the age range of the clients but who were perceived by the adolescents as optimally similar for communication purposes. These near-peers ideally combined a high degree of trustworthiness-credibility with some competence-credibility. In other words, the providers were trustworthy to adolescents since they were just beyond adolescence but still of the "youth generation." At the same time, the near-peers were perceived as competent because they were somewhat older, presumably more experienced, and had training in peer counseling and in any necessary technical information. Conventional service providers usually can only offer competence-credibility; the result is that frequently their very accurate and useful information is not accepted by clients as trustworthy or "safe." Most peers, however, present the opposite problem (and adolescents are aware of it). While sex-related information coming from peers is perceived as very trustworthy, it may be inaccurate. Contraceptive "teen clinics" provide a happy medium; thus the San Francisco Teen Clinic provided a model that is now widely copied throughout the United States.

The two programs described below exemplify some of the more recent trends in peer counseling.

1. The program of the Planned Parenthood Association of Marin County (in San Rafael, California) recruited and trained sixteen peer counselors in three high schools in 1976–77. The counselors were recruited from classes in human sexuality, trained in six evening sessions (on listening skills, sexuality, and contraceptive methods) and then asked to lead—under adult supervision—teen "rap" sessions at the local Planned Parenthood clinic. No counseling was formally scheduled in the high schools, but a great deal of informal counseling occurred. Although they received academic credit, the peer counselors were unpaid. One problem arose: some of the peer counselors dropped out of the program, so that by the end of the 1976–77 school year, only ten of the original sixteen counselors remained.

2. At George Mason High School in Falls Church, Virginia, twenty-four peer counselors were trained in 1975–77. The counselors were recruited from courses in human relations and sex education. Prior to beginning their work, they completed two of three courses in peer counseling, including a peer counseling practicum. The counseling itself, which earns school credit, covers a wide range of topics such as study skills and personal adjustment problems, as well as contraception and sexuality. The relatively small size of the school (750 students in grades seven through twelve) may be one of the factors in the relative success of the program.

Many Planned Parenthood clinics that use peer counseling as part of their services establish special times of the day and special rooms to accommodate adolescent clients. Such teen clinics are usually located in comfortable, brightly decorated settings where the adolescent clients sit on the floor or in bean-bag chairs. The near-peer counselor dresses informally and uses language that is understood by the teenaged clients. Their services, however, are usually limited to contraceptive information and services, typically provided in a one- or two-visit interaction with the adolescent client. Individuals who come to Planned Parenthood teen clinics become more aware of their need to be informed about contraception (many think they may be pregnant), and perhaps they then pass along this information to friends in school and elsewhere.

Perhaps fairly typical of university-level peer counseling programs is the Human Sexuality Information and Counseling Service at the University of North Carolina.[50] This service emphasizes preventative counseling and support in "subacute" crises, and referral to appropriate professional services.

Peer counseling provides information about sexuality as well as a context (social and psychological) for processing and exploring it for personal relevance. Additionally, peer counseling has often changed sexuality from a problem-centered medical issue to an informational and self-exploratory topic. Peer counseling often serves individuals who are not directly experiencing problems, but who nonetheless need information or value clarification in an informal setting.

Intervention designed to help adolescents cope with sex-related problems resorted by fortuitous accident to using peer group dynamics. The key to reaching this age group with counseling, information, advice, and opportunities to talk seems to be an egalitarian relationship. Social service agencies clearly cannot expect to be consulted on sex-related matters by teenagers except when dire necessity (such as pregnancy) so dictates. The initiative to reach out and offer help and true concern must come from the agency or institution. Moreover, in order to achieve success, the agency must enter as far as possible the adolescent milieu.

CONCLUSIONS

This chapter has synthesized what is known, and should be studied, about peer communication in learning about sexuality. We suggest the following implications.

1. The dependence of adolescents upon peer communication means that the quality of information about sexuality thus provided is often inaccurate. An adolescent's peers are usually not better informed about sexually related matters than is the individual. Many adolescents have inadequate, inaccurate, or superficial knowledge about contraception, reproduction, and other aspects of sex informa-

tion. It is here that educational intervention is needed to provide more accurate and more comprehensive information about sexuality.

2. Much peer communication about sexuality flows through networks that are homophilous and interlocking, and thus are of limited value for transmitting new information. But there is some penetration of adolescent networks by near-peers, parents, and other adults, certain mass media, and some social service agencies. Some intervention organizations use peer counseling as a means of providing sex information to adolescents through near-peers. Fellow adolescents, who may be somewhat older and have more specialized training than the adolescent client, are perceived as possessing high trustworthiness-credibility and some competence-credibility by the users of agency services.

3. Much research remains to be done to explore the nature of peer communication with respect to sexuality. The most promising approach may be communication network analysis, a technique that can investigate the interactions among people rather than specific individuals' characteristics. By understanding the types of relationships, their functions, and the communication content common to peer groups, we may learn more about the nature of sexual learning for peers. In fact, a number of studies on communication networks have recently been completed or are underway.[51]

NOTES TO CHAPTER 6

1. J. Piaget, *The Language of Thought of the Child* (New York: Harcourt, Brace, 1926).
2. D.C. Dunphy, *Cliques, Crowds and Gangs* (Melbourne: F.W. Cheshire, 1969).
3. Janet Lever, "Sex Differences in the Complexity of Children's Play," *American Sociological Review* 43 (1978): 471–482.
4. Ibid., p. 472.
5. Ibid., p. 481.
6. Stanley Aronowitz, *False Promises: The Shaping of American Working Class Consciousness* (New York: McGraw-Hill, 1973).
7. Ibid., p. 64.
8. Iona Opie and Peter Opie, *The Lore and Language of Schoolchildren* (Oxford: Clarendon Press, 1959).
9. Jacob Lomranz et al. "Children's Personal Space as a Function of Age and Sex," *Developmental Psychology* 11 (1975): 541–545.
10. Nancy M. Henley, *Body Politics: Power, Sex, and Nonverbal Communication* (Englewood Cliffs, N.J.: Prentice-Hall, 1975), p. 36.
11. Ibid.

12. E.P. Hollander and J.E. Marcia, "Parental Determinants of Peer Orientation and Self-Orientation Among Pre-Adolescents," *Developmental Psychology* 2 (1970): 292–302.

13. James W. Maddock, "Sex in Adolescence: Its Meaning and Future," *Adolescence* 8 (1973): 325–342.

14. Ibid., p. 331.

15. Stuart Ewen, *Captains of Consciousness: Advertising and the Social Roots of the Consumer Culture* (New York: McGraw-Hill, 1976).

16. U.S., President's Science Advisory Committee, Panel on Youth, *Youth: Transition to Adulthood* (Chicago: University of Chicago Press, 1974), p. 112.

17. Robert S. Lynd and Helen M. Lynd, *Middletown: A Study in Contemporary American Culture* (New York: Harcourt, Brace, 1929); August Hollingshead, *Elmtown's Youth: The Impact of Social Class on Adolescents* (New York: Wiley, 1949).

18. Joseph Kett, *Rites of Passage* (New York: Basic Books, 1977), p. 269.

19. U.S., President's Science Advisory Committee, *Youth.*, p. 2.

20. Ibid., p. 2.

21. Ibid.

22. D.P. Ausubel et al., *Theory and Problems of Adolescent Development* (New York: Grune and Stratton, 1954).

23. Ibid., p. 344.

24. John Johnstone, "Social Integration and Mass Media Use Among Adolescents: A Case Study," in *The Uses of Mass Communications*, eds. Elihu Katz and Jay Blumer (Beverly Hills: Sage Publications, 1974), pp. 35–47; P.A. Witty, "Television and the High School Student," *Education* 72 (1951): 242–251; M.W. Riley and J.W. Riley, "A Sociological Approach to Communications Research," *Public Opinion Quarterly* 15 (1951): 444–460; Joseph R. Dominick, "The Portable Friend: Radio," *Journal of Broadcasting* 18 (1974): 161–170.

25. James S. Coleman, *The Adolescent Society: The Social Life of the Teenager and its Impact on Education* (New York: Free Press, 1961).

26. Gerald D. Suttles, *The Social Order of the Slum: Ethnicity and Territory in the Inner City* (Chicago: University of Chicago Press, 1968).

27. Patricia Miller and William Simon, "Adolescent Sexual Behavior: Content and Change," *Social Problems* 22 (1974): 58–75.

28. "The really meaningful significant others in the adolescent's life space are his fellow adolescents." Ernest J. Campbell, "Adolescent Socialization," in *Handbook of Socialization Theory and Research*, ed. David A. Goslin (Chicago: Rand McNally, 1968).

Further evidence for the importance of peer influence on adolescents is provided by Glenn V. Ramsey, "Sex Information, Attitudes, and Behavior," in *The Adolescent*, ed. Jerome M. Seidman (New York: Holt, Rhinehart, and Winston, 1960); Graham B. Spanner, "Sources of Sex Information and Premarital Sexual Behavior," *Journal of Sex Research* 13 (1977): 73–88; Luther B. Otto, "Girl Friends as Significant-Others: Their Influence on Young Men's Career Aspirations and Achievements," *Sociometry* 40 (1977): 287–293.

29. It is important to realize that each of these socializing influences interacts with others, so that it is an oversimplification to identify a "most impor-

tant" influence. See Richard M. Perloff and Michael E. Lamb, "The Development of Gender Roles: An Integrative Life-Span Perspective" (Department of Psychology paper, University of Wisconsin).

30. Ramsey, "Sex Information."

31. Robert C. Sorensen, *Adolescent Sexuality in Contemporary America* (New York: World, 1973).

32. Jack O. Balswick and James W. Balkwell, "Self-Disclosure to Same- and Opposite-Sex Parents: An Empirical Test of Insights from Role Theory," *Sociometry* 40 (1977): 282–286.

33. Perloff and Lamb, "Development of Gender Roles."

34. Albert Bandura, "The Stormy Decade: Fact or Fiction?" *Psychology in the Schools* 1 (1964): 224–231.

35. Ramsey, "Sex Information."

36. Miller and Simon, "Adolescent Sexual Behavior."

37. Frank F. Furstenburg, Jr., *Unplanned Parenthood: The Social Consequences of Teenage Childbearing* (New York: Free Press, 1976); John F. Kantner and Melvin Zelnick, "Sexual Experiences of Young Unmarried Women in the United States," *Family Planning Perspectives* 4 (1972): 9–18.

38. Ramsey, "Sex Information."

39. J. Roy Hopkins, "Sexual Behavior in Adolescence," *Journal of Social Issues* 33 (1977): 67–85.

40. *Homophily* is the degree to which pairs of individuals who interact have similar beliefs, knowledge, socioeconomic status, and the like. In contrast, *heterophily* is the degree to which pairs of interacting individuals are unlike. See Everett M. Rogers, *Communication Strategies for Family Planning* (New York: Free Press, 1973).

41. *Personal communication networks* consist of interconnected individuals who are linked by patterned communication flows to a given individual. A personal network is anchored on a single individual, and consists of the other individuals with whom he or she interacts directly and indirectly.

42. Spanner ("Sources of Sex Information") found that a national sample of college students (mostly in late adolescence or postadolescence) reported that independent reading about sex matters was almost as important as peer communication about this topic.

43. Otto, "Girl Friends."

44. Such sex differences for later childhood are documented by M.F. Waldrop and C.F. Halverson, "Intensive and Extensive Peer Behavior: Longitudinal and Cross-Sectional Analyses," *Child Development* 46 (1975): 19–26. However, there is not much evidence on whether this pattern of different personal networks for males and females continues in adolescence. Clique structures for adolescent females are more complex and interlocking than are those for boys. See Coleman, *Adolescent Society*.

45. Guttmacher Institute, *11 Million Teenagers: What Can Be Done About the Epidemic of Adolescent Pregnancies in the United States* (New York: Planned Parenthood Federation of America, 1976).

46. Michael Baizerman, "Toward Analysis of the Relations Among the Youth Counterculture, Telephone Hotlines, and Anonymity," *Journal of Youth and Adolescence* 3 (1974): 293–305.

47. Ibid.

48. Graham B. Spanier, "Sexualization and Premarital Sexual Behavior," *Family Coordinator* 24 (1975): 33–41; Planned Parenthood San Francisco/Alameda County, Unpublished statistics, 1975–76; Joseph P. Frolkis, "Adolescent Pregnancy: A Critical Review" (paper for The Brush Foundation, 1977).

49. Much adult resistance to providing contraceptive information and services to adolescents is based on the assumption that the availability of contraceptives leads to sexual activity; however, much evidence is available that adolescents typically seek contraceptives after having had sex for an average of about one year. In fact, most new adolescent patients at family planning clinics come because they think they are pregnant. See Urban and Rural Systems Associates, *Improving Family Planning Services: Executive Summary*, (Final Report, Contract HEW 05–74–304, U.S. Department of Health, Education and Welfare), 1972; Diane Settage, "Sexual Experiences of Younger Teenage Girls Seeking Contraceptive Assitance for the First Time," *Family Planning Perspectives* 5 (1973): 223–226.

50. Bruce A. Baldwin and Robert E. Staub, "Peers and Human Sexuality Outreach Educators in the Campus Community," *Journal of the American College Health Association* 24 (1976): 290–293.

51. Everett M. Rogers and D. Lawrence Kincaid, *Communication Networks: A New Paradigm for Research* (New York: Free Press, in press).

Chapter 7

Social Services and Sexual Learning

Mary Jo Bane with
*Steven A. Holt**

INTRODUCTION

The family, the school, and the peer group are the most important institutions in the lives of most American children. Nonetheless, children also encounter a number of other institutions that can have important effects on various aspects of their social and sexual development. These institutions are the social welfare organizations of the public and private sector—churches, voluntary groups, health services, social service agencies, public assistance offices, child welfare and protective services—that are increasingly important parts of the environments of families and children.

This chapter examines some of the service-providing institutions from the perspective of their impact on what is broadly defined as the sexual development of children. This perspective is not as far-fetched as it might seem at first glance. Some services, particularly health care and youth organizations, have obvious and direct effects on sexuality and sexual behavior. Others have more subtle impact through the assumptions that they embody and project about appropriate family and life-style options, appropriate roles of men and

*As a writer in the field of social planning, Mary Jo Bane is best known for her book *Here to Stay: American Family in the Twentieth Century*. Dr. Bane is Associate Professor of Education at the Harvard Graduate School of Education, Associate Director of the Joint Center for Urban Studies at Harvard and MIT, and consulting editor for the *American Journal of Sociology*.

Steven A. Holt is Director of the Project on Human Sexual Development, has been integrally involved in all phases of the Project's work, and is coauthor of "Television and Human Sexuality" and "Parent/Child Communication about Sexuality."

women, social and moral responsibility, perceptions of the body, and displays of affection.

This chapter will first describe the institutions that make up the social welfare system and will identify those that are most likely to be important in the lives of children. It will then describe the kinds of messages these institutions (considered as a system) may be transmitting to children on the various aspects of sexual learning. Finally, it will look in more detail at the specific content of messages sent by four institutions.

I. SOCIAL WELFARE INSTITUTIONS

The social welfare system embodies the concern of the society as a whole for the health and well-being of its member individuals and families. It consists of public and private voluntary ("charitable") programs and agencies that provide income, employment and training, housing, education, recreation, health care, and personal social services to people who need them. The last category—personal social services—includes such services as casework, counseling, day care, family planning, information and referral, and child welfare services.

The growing importance of the social welfare sector of American society can be documented in a number of ways. Naturally, public and private expenditures on social welfare (including income maintenance, education, and health) have grown from $78.7 billion in 1960 to $395.9 billion in 1975—a change from 15.8 percent to 27.5 percent of the Gross National Product.[1] The number of programs and agencies providing services is now very large. For example, the *Catalog of Federal Domestic Assistance* for fiscal year 1978 lists 180 different federal programs under the heading "Income Security and Social Services."[2] In a city such as Boston the listing "Social Service Agencies" in the Yellow Pages takes up three pages of small print.

Such a large sector of the society must be affecting children, directly and indirectly, in important ways. Many children go with their parents (or see their neighbors go) to the welfare office to apply for food stamps or public assistance. Children hear adults talk about the forms, the rules and the questions. Children know that welfare is an important part of their lives. More privileged children, with no direct experience of welfare, know about it from TV and newspapers. They, too, learn something about who gets welfare, why some people get it, and what rules the recipients must follow.

Other services can also become important parts of children's lives. Many children belong to church and youth recreation groups. Many are seen or tested by school psychologists; some are intensely involved in counseling, therapy, or remedial programs. A smaller number of children become wards of the public welfare system in foster homes or institutions. Again, however, even those children with no direct ties to the service-providing system commonly know of its existence and something of its assumptions and rules through the experience and talk of family and friends and through the media.

Because data collection in the social service sector is so poor, it is impossible to find out precisely how many children come in contact with the various public and private agencies that provide services. Thus, speculations about impact and importance can be only that. We suggest, however, that the following agencies are probably the most important in the lives of children.

Services associated with schools. Nearly all American children between the ages of six and sixteen attend schools. Schools today do much more than simply educate. Among other services, they provide counseling and guidance, some health care and sex education, sports, recreation programs, and lunch. The new federal legislation that requires school districts to provide free and appropriate education and services to the handicapped is likely to increase dramatically the role of the schools in providing services directly to the 10 or 15 percent of children with some identifiable handicap, and indirectly to others. Simply because of their pervasiveness, schools are probably the most important service-providing institution with which children come in contact.

Services associated with churches. Approximately 60 percent of the American population claims membership in a church.[3] Like schools, therefore, churches broadly cover the population, including the population of children. Just as schools do not limit their activities to academic training, churches do not limit their activities to worship; many of them provide counseling, recreation, and other social programs.

Youth organizations. Large numbers of children have some contact with one or another of the national youth organizations: Boy and Girl Scouts, Boys and Girls Clubs, 4–H Clubs, Little League, Pop Warner League, and others. For example, in 1976 the Boy Scouts reported 4.9 million members while the Girl Scouts claimed 3.2 million members. Of the Girl Scouts, 1.2 million are Brownies, aged seven and eight; this is about 35 percent of the population in that age group.[4] These organizations sponsor recreational and educational activities, many of which deal explicitly with the process of becoming physically and socially adult.

Health care and family planning. In recent years health care has grown to be one of the largest industries in the United States. Although much of this care is delivered in private physicians' offices (and thus falls outside our definition of the social welfare system), other settings are relevant to this discussion. Most children will have some experience with hospitals, out-patient clinics, community health centers, or family-planning clinics. The few children who do not have direct contact with any of these may become involved via a family member or friend.

Public cash assistance programs. The major cash assistance programs benefit large numbers of children. In 1976, recipients included 2.9 million children of deceased workers, 1.5 million children of disabled workers, and 0.7 million children of retired workers under Social Security, 7.9 million children under Aid to Families with Dependent Children (AFDC), and some proportion of the 18 million food stamps recipients.[5] In addition, many children have close relatives, especially grandparents, who receive Social Security benefits. Children who receive cash assistance are also likely to have contact with some of the social services authorized under Title XX of the Social Security Act (grants to states for social services, which include day-care, family planning, case work, home help, and other services).

Child welfare, protective, and foster care services. The number of children who come into the child welfare system is not as large as those affected by the other institutions listed here. Nonetheless, for those who do come under state protection—and especially for those placed in foster care—the system has dramatic and far-reaching effects on their lives.

II. MESSAGES ABOUT SEXUALITY

Social welfare institutions, like other societal institutions, send "messages" to the people who come in contact with them. Some of the messages are explicit: group leaders, coaches, or social workers actually express certain messages to their clients. Other messages are implicit in the structure and underlying assumptions of the institutions. The messages sent by the social welfare system are delivered, we believe, in five important areas of sexual learning: values and standards, responsibility, gender roles, family and life-style options, and images of the body.

Message Areas

Values and Standards. What is right and wrong, acceptable and unacceptable, or appropriate and inappropriate with regard to sexual attitudes and behavior? What is all right to think about, talk about, do? Social services can give answers to these questions in many ways: in what counselors and leaders say, and in what they permit groups to talk about; in how teachers, pastors, and health professionals deal with birth control, abortion, and venereal disease; in how recreation programs define appropriate behavior for boys and girls; in how public assistance programs deal with illegitimacy and cohabitation.

Responsibility. For whom am I responsible—physically, emotionally, and financially? Who is responsible for me? These questions are obviously crucial in the development of mature sexuality, love, and care. Our culture sends many messages on responsibility. Some imply that people are responsible only for themselves; others imply that the government is ultimately responsible; and yet others indicate that people who live with, are related to, or care for each other are also responsible for each other's well-being. The social welfare system, along with the other institutions that children encounter, embodies assumptions about responsibility that are important to identify and understand in specific contexts. What do family planning services imply about male and female responsibilities to each other and to their children? Where do counselors locate responsibility for personal problems? What does the cash assistance system assume about the financial responsibility of family members for each other and about the responsibility of the community for its citizens? How do child welfare workers dealing with abuse and neglect cases see family and state responsibility?

Gender Roles. What is appropriate behavior for men and women, boys and girls? This is another obviously important question for sexual learning, and another area in which social welfare institutions may be sending a variety of messages. Who is allowed or encouraged to participate in sports, recreation programs, and nonacademic learning experiences? Do counselors or case workers say or imply that boys and girls are capable of, or suitable for, different jobs? Are male and female parents assumed to have different kinds of responsibility for children? Do the organizations themselves exhibit a sexual hierarchy?

Family and Life-Style Options. What are acceptable forms of family and nonfamily life? Is it okay not to marry, not to have children, to live alone, to cohabit, to divorce, to love a person of the same sex? A variety of programs give answers to these questions. How do child welfare workers evaluate different kinds of living arrangements? What do eligibility rules for public assistance imply? What do health professionals imply about marrying, having children, living in various kinds of arrangements?

Images of the Body. Sports and recreation and educational activities of youth organizations can raise a number of questions—and provide their own implicit or explicit answers—about physical development and perceptions of the body. Is a strong, healthy body to be enjoyed for itself, or for its value in competition? Is physical activity perceived as a substitute for sex, thus degrading both? Are parts of the body to be kept covered and hidden from members of the same or other sex? Can one be comfortable with one's physical strengths and limitations?

Mixed Messages and the Importance of Federal Funding

The bulk of this chapter looks at the messages conveyed in these various areas by some of the different institutions of the social welfare system. An underlying theme of the analysis is that the system is probably sending out mixed sets of messages in most areas. We believe this results from inherent tensions in the system between various constituencies: among families with different values, between professional and community reference groups, and between national and local norms.

Both private and public service agencies are responsive to several different constituencies that pull the agencies in different directions. Public agencies must attend to the taxpayers, and private agencies must pay heed to their donors. These funding groups may have values and standards that differ from the agencies' clients, or they may expect agencies to hold their clients to more rigid standards of behavior and propriety than are expected of those who do not receive "charity." Service workers may thus feel compelled to encourage or demand adherence by their clients to the standards of the community—however such standards are defined.

The working- or middle-class background of social welfare professionals may reinforce this stance. At the same time, however, the national social work profession makes up an important reference group that embodies the usually more liberal standards of the intellectual elite. Social welfare workers may often be caught in internal conflicts as well—with volunteers or paraprofessionals holding one set of values and attitudes, ordinary case workers another, and the professional administrators yet a third.

The tensions are complicated by the fact that most private and public agencies now receive funds from several different sources. They may receive funds from private donors who must be propitiated, from local governments whose taxpayers must be placated, and from the federal government. Federal money, of course, brings with it regulations and standards designed at the national level, usually by the professional elite. With the increasing importance of federal funding, conflicting expectations from different constituencies have become an important feature of agency life.

Nationally, between 1960 and 1975, the public share of what was broadly defined as social welfare spending rose from 65.3 percent to 73.1 percent of total spending. Within the public sphere, federal (as opposed to state and local) spending rose from 47.8 percent to 58.4 percent.[6] The rise in the federal share of social welfare spending is accounted for mostly by federal social insurance, cash assistance, and health programs. Federal funding in other areas is also important, however, since it brings a large number of service activities into the sphere of federal regulations.

Most school districts, for example, receive some federal money and must obey federal prohibitions against sex discrimination (Title IX of the Education Amendments of 1972). Sports, recreation programs, career guidance, and other services must be provided equally and on a nondiscriminatory basis to boys and girls. Youth organizations that use public facilities or receive public funds are also bound, at least to some extent, by federal antidiscrimination regulations. Such relationships have profound implications for the messages schools and youth organizations deliver about appropriate gender roles. Teachers and leaders may find themselves caught between different sets of demands, and children may be hearing several different messages. This is not necessarily bad, of course, since it may provide children with opportunities to develop their own sets of standards and find reinforcement for them. It may also mute the effect of some of the rigid gender-role expectations that many schools still teach.

Other examples arise from the fact that funds from both Title XX of the federal Social Security Act (grants to states for social services) and the Comprehensive Employment and Training Act (CETA) (salaries for workers) are now important sources of income for a large number of social service agencies. These agencies are, in turn, subject to a variety of federal regulations. Ironically, CETA may also have its own impact on the social welfare system. To the extent that the purposes of CETA (to provide jobs for the unemployed) are honored in the hiring of workers (and they are not always), CETA workers are unlikely to be professionals. Their presence in social service agencies in large numbers may thus be tilting the balance away from national professional standards back toward local community norms—contributing yet another ingredient to the mix of messages.

Child welfare services also receive a substantial amount of federal money (Title IV-B of the Social Security Act). Federal regulations in this area cover case management, the quality of services, and, perhaps most important, the rights of parents to "due process of law" before children can be removed from their homes. These regulations have an important impact on the degree to which services are professionalized and on the allocation between parents and the state of responsibility for children. Social workers and social welfare institutions are taught to place the welfare of the child above all other considerations. They often come to see themselves as protectors of children against parents who are perceived as incompetent. Parents are often denigrated or, at best, ignored. The new federal presence, however, both through regulations and through case law, emphasizes the rights of parents and presents a challenge to the traditional orientation of the social work profession. Federal funding is thus setting up a new set of norms (about power, responsibility, and appropriate behavior) that coexist, often uneasily, with the old.

Health and family planning services receive federal money directly through Medicaid. They too become subject to federal regulations and federal laws, which now mandate nondiscrimination and confidentiality.

Finally, the federal cash assistance system has well-specified eligibility rules, definitions of responsibility (defining when, for example, a man living in a household is required to contribute to the financial support of children in the household), and requirements for fair procedures when benefits are granted and withdrawn. These are much more liberal than the standards of many communities and may be moving welfare departments in that direction. (This may also, of course, be undermining community sympathy for welfare recipients and support for welfare programs.)

The importance of increased federal funding and federal regulation of the social welfare system cannot be overemphasized. Professionalism, proceduralism, and nondiscrimination—introduced along with the federal presence—are coming into tension with the local, informal, and particularistic norms of much of the social welfare tradition and of many users of the social services system.

That agencies repond to many different constituencies; the fact that different values are in conflict with each other within agencies (and often within individual professionals) means that agencies seldom deliver coherent or consistent messages in many of the important areas of sexual learning. Conflicting messages may cause tension and confusion. On the other hand, they may offer an opportunity for those who receive their messages—be they clients, workers or observers—to expand and clarify their own options. Confusion in the messages delivered by the service system may increase the salience of messages coming from other institutions. It may also provide an opportunity for people to consider a number of possibilities in forming their own notions of the different aspects of sexuality.

III. SPECIFIC INSTITUTIONS

The complex set of messages about sexuality that social service institutions deliver is best approached by analyzing specific institutions. The following sections will highlight what seem to be the more important messages of four important institutions that affect children. (Despite their importance, neither schools nor churches are dealt with here because they are examined elsewhere. See Chapters 5 and 8.) I will look at the cash assistance system, health care institutions, youth organizations (using the Scouts as an example), and the child welfare system.

Public Cash Assistance Programs

In terms of dollars, public cash assistance and quasi-cash assistance programs are by far the largest social welfare programs that affect children. Current programs include the following:

1. Old Age, Survivors, Disability, and Health Insurance (Social Security): 5 million beneficiaries in 1976 were children of deceased, retired, and disabled workers [7]

2. Aid to Families with Dependent Children (AFDC): 7.9 million children received some benefits in 1976; average monthly cash benefit to families was $236 [8]

3. Food Stamps: In 1976 about 18 million Americans participated in the Food Stamp Program. An unreported but probably large proportion were children. The program overlaps partially but not

completely with public assistance; about 40 percent of food stamp users are not public assistance recipients[9]

4. Housing Assistance: The federal government provides subsidized public housing, rent subsidies, and mortgage insurance to low-income families under a number of Housing and Urban Development programs. Data are not reported in such a way as to allow estimates of the number of beneficiaries. Budget outlays, however, indicate that housing assistance is substantial and widespread. Because both food stamps and housing assistance are essentially substitutable for cash, they can be considered forms of cash assistance.

These figures indicate that large numbers of children live in families that receive some form of public cash assistance. Because turnover in all the programs is very high, many more children will be beneficiaries of cash assistance at some point during their childhood than appear on the rolls at any one time. In addition, many children will know about the cash assistance system because relatives or neighbors receive aid, even if their own families do not.

Cash assistance beneficiaries remain a minority, though a substantial minority, of children. Nonrecipients, however, also receive messages from the system. They know of the existence of welfare, know that certain kinds of people receive it, and may know some of the ways in which welfare works. The dramatic growth of the welfare system over the last fifteen years has made it a pervasive feature of American life.

The way in which welfare programs are administered, their eligibility requirements and governing regulations, and the interactions that occur between welfare workers and clients transmit messages about responsibility, gender roles, and family and life-style options. Their messages are not, of course, directed explicitly at children, and it is possible that children are in fact protected from their impact. Social Security, for example, is a relatively unintrusive program, involving a straightforward eligibility determination and no casework. Children may not even know that part of the family income is coming from Social Security, and if they do, may not perceive any impact on family life other than the simple fact of having more money.

AFDC, on the other hand, is likely to be perceived as a substantial factor in its clients' lives. Even if children do not accompany their mothers to the welfare office, they are likely to be present when caseworkers visit. Moreover, the mother's interaction with the caseworker, and the process of eligibility determination may be sufficiently difficult and salient that she relates her experience and her complaints to the children. At least for AFDC, then, it seems useful to examine the messages in order to see what children might learn from the eligibility requirements and administrative practices of the program.

AFDC is a cooperative federal-state program, with the states retaining control over most program details. This results in considerable variability from state to state. The material presented here is based mainly on the federal requirements.[10]

Messages about Responsibility. Messages about responsibility are conveyed by the mere existence of an AFDC program, by what is counted as income in the process of eligibility determination, by regulations on relative responsibility and their enforcement or lack thereof, and by the new Child Support Enforcement Program.

The general society transmits in many ways quite strong messages that people are responsible for their own support and that parents are responsible for supporting their children. The existence of an AFDC program modifies that message somewhat: it says the government will provide some financial support for needy families with children. Though nonrecipients often perceive welfare as charity, and though historically aid was often administered like charity, AFDC is now available as a matter of right. Eligible applicants are entitled to receive support, and assistance cannot be terminated without prior notice and a fair hearing (by decision of the U.S. Supreme Court in *Goldberg* v. *Kelly* [1970]).

Government support comes at a price, of course, paid mostly in self-respect and privacy. AFDC requires a means test, and is available only to those who can prove that their incomes are very low. It is thus unlike Social Security, eligibility for which is based on previous contributions. Because both eligibility and benefit level under AFDC are determined by "need," caseworkers make detailed inquiries into the sources of income, living arrangements, and consumption patterns of their clients. The message is that the poor are different from other people, and that they are less worthy of having their honesty and privacy respected. Parents may be devalued in the eyes of their children; they may be seen as less responsible and less worthy of respect.

Another message about responsibility is sent by the Child Support Enforcement Program. Under legislation passed by Congress in 1975, AFDC clients (almost all of whom are women) are required to report the name of the absent parent or parents of their children to the welfare worker, and to sign over rights to child-support payments to the welfare department. State and county welfare departments are required to track down and collect child-support payments from absent parents, with the aid of the federal Parent Locater Service. This service is supposedly available to nonrecipients of AFDC as well—child-support payments from absent parents may keep families off welfare—but many states restrict their efforts to welfare clients.

The Child Support Enforcement Program embodies the assumptions that parents are financially responsible for children, but that poor parents are likely to shirk that responsibility. Unlike middle-class parents, they must be forced by the government to meet their responsibilities. Poor children may learn from this that fathers (almost all absent parents are male) are expected to be financially irresponsible. The pressure from the welfare department, and fathers' attempts to avoid caseworkers, may make the fathers emotionally irresponsible as well. The burden of physical care and emotional support rests on the mother, who is not expected to provide financial support. The two kinds of responsibility are borne by different people and are seen as separate.

Two sets of assumptions about responsibility would thus seem to characterize the AFDC program. First, the program assumes that responsibility for the care of children rests and should rest with one or both of their parents. Financial responsibility rests primarily on the parent with whom the child lives, secondarily with the absent parent, and finally with the government. Responsibilities of grandparents and adult siblings are defined by some states but enforced only sporadically. Partners and spouses of parents with custody of children are not considered responsible for the children as long as such partners have not formally adopted the children. Responsibility, in short, is quite narrowly defined. Second, the AFDC program assumes that the financial responsibilities of absent parents, who are almost always men, have to be enforced. The men must be located and made to pay. This assumption reflects a practical fact—that is, the poor record of fathers, including middle-class fathers, in paying child-support and alimony. But the perception that fathers are reluctant supporters of children, requiring coercion by a government agency, is reinforced by the program.

Messages about Gender Roles. Differences in responsibility, discussed above, are obviously part of the message that AFDC transmits about gender roles. But there are other messages as well. Most legislation and regulations on AFDC are now written in sex-neutral language—speaking of "parents" and "caretakers," and defining rights and obligations by function rather than sex. Provisions covering work requirements, however, still make some distinctions on the basis of sex, and imply that men are expected to work and support their families financially, while women are expected to be supported as they care for home and children. These expectations may be reinforcing cultural norms about appropriate roles for men and women.

Messages about Family and Life-Style Options. Much of what has been said about responsibility and gender roles also has implications for learning about family and life-style options. Notions of financial responsibility as narrowly defined as those of AFDC would seem to imply an expectation of small, individualized, tightly defined households. In general, the AFDC benefit structure—for those persistent enough to figure it out—also encourages small, separate families: two small households receive higher total benefits than one large household.

Efforts are now being made to reform the benefit structure so as not to discourage the formation of larger households—for example, single parents living with their parents. The problem is very complicated, however, since large households are believed to benefit from economies of scale (especially in housing costs) and thus to need less per capita income than smaller households.

It is further complicated by disagreements over family responsibility. Some argue, for example, that a single mother ought to be able to collect welfare on her own, even if she is living with her own parents, and that the grandparents' income ought not to be considered in figuring the mother's benefits. This provision would remove the incentive for the mother to establish her own household in order to collect welfare. Furthermore, the arrangement would make the emotional and social support of the grandparents' home available to the mother and her children. Others argue that such a rule would imply approval of grandparents withholding financial support from their children and grandchildren. Such an implication, they argue, violates our notions of responsibility and weakens family ties. Welfare has not resolved this dilemma.

The AFDC program has also faced, not entirely successfully, the issues of cohabitation and remarriage. In the past, both were discouraged. Cohabitation was explicitly discouraged through denial of benefits when there was a man in the house and through midnight searches by welfare workers. Remarriage, while officially encouraged, also led to a cut-off of benefits—hardly an incentive. After the U.S. Supreme Court declared that unannounced searches for men were unconstitutional invasions of privacy, the sanctions against cohabitation were reduced to social worker finger-wagging. In other important decisions, the courts decided that neither cohabitation nor remarriage necessarily meant that the resources of the partner were available to the children. Benefits could not, therefore, be denied or reduced unless the children were officially adopted by the new partner.

These rulings have made the AFDC program more nearly neutral than it was with regard to family and life-style options, and probably more nearly neutral than the standards of the society as a whole. Court decisions and revisions of regulations are continuing the trend. It is not known whether local workers have fully accepted these rules in practice. What is clear is that the old moralistic standards of welfare work have been successfully challenged in the courts. Workers must now pay at least lip service to the legitimacy of a plurality of life-style options.

The very existence of separate AFDC and Social Security programs suggests, however, that some options are more legitimate than others. Social Security (to the extent it affects children) supports children of disabled and retired workers, and widows and orphans of covered workers. These are the "deserving poor" for whom eligibility is determined almost automatically and for whom benefits are set at a reasonable level related to previous earnings. In contrast, divorced, separated, and unwed mothers—historically looked down upon—are subject to the now humane but still intrusive and demeaning practices of AFDC. It must seem to children of AFDC recipients that their parents have done something wrong in divorcing or separating and in being poor that makes it legitimate for other people to pry into their affairs and to require them to live on very little money.

Health Care

Over the last two decades—and especially since government funding of Medicare and Medicaid—the health business has become a huge growth industry, involving large pharmaceutical, hospital supply, construction, and insurance companies. In the 1960s, the total amount spent for health care in the United States doubled, reaching $62 billion in 1969. The figure was expected to climb to over $94 billion by 1975. Instead, the total reached $104.2 billion in fiscal 1974, to make the health field the nation's largest industry both in terms of dollar volume and number of people employed.[11]

All preadolescent children are likely to have at least some contact with health professionals. At least one out of every ten children is expected to spend some time in a hospital this year—for emergency room care, clinical tests, or surgery.[12] Apart from this exposure, children may visit a doctor, health clinic, or school nurse for immunizations, check-ups, first aid for accidents, and treatments for diseases. Children who are not themselves involved with health professionals may be affected by the interactions of family members and friends with these systems and by mass media portrayals of medical treatment.

Describing the attitudes and behavior of health professionals is extremely difficult because of the diversity of settings and the variety of specialists now working in the field. Hospitals and clinics vary according to physical size, patient-staff ratio, and liberality of outlook. The child may have primary contact with male or female family doctors, specialists, medical technicians, nurse practitioners, or a combination of all of these. Moreover, the child may be treated on an adult ward or in a children's hospital and may have either a brief encounter or an extended treatment.

Relatively few of the encounters that children have with health professionals are likely to be over explicit sexual issues. Family planning, pregnancy, and sexual behavior will probably not become subjects for professional consultation until children are older. Sexual learning, therefore, will occur from less direct messages—particularly from the attitudes held by professionals, from assumptions underlying government-sponsored programs such as Medicare and Medicaid, and from the physical settings of institutions.

As with other social service institutions, the health care system, because of its vastness, will inevitably send a mixture of messages about sexuality to children. Messages from the primary physician, for instance, may differ significantly from those of a lab technician, nurse practitioner, or receptionist, from the hospital setting itself, or from the qualifying regulations for government aid.

Messages about Values and Standards. An important statement about the medical profession's values with regard to sexuality can be seen in current approaches to medical training. A 1977 survey indicated that although 81 percent of medical schools have "some training in human sexuality," many courses are electives or parts of other courses (obstetrics and gynecology or psychiatry, for example).[13] About 50 percent of U.S. medical schools have no special distinctive courses in human sexuality. Despite enormous advances in the teaching of human sexuality since 1960 (when only three medical schools offered any formal instruction in the subject), sexuality—as a developmental issue, apart from reproduction—still has little place in client-doctor discussions.[14] This is true despite the fact that most people report that they believe doctors are among the most accurate (if underused) sources of information about sexuality.[15]

In addition to sparseness of training in human sexuality, the modern trend of medicine toward training specialization results in compartmentalization of health care. Treatment is likely to diagnose quite narrowly and to refer the patient to a physician skilled in a very particular science. Sexual matters, therefore, would not be routinely handled by doctors other than those trained in psychiatry, or therapy, or some other area that is believed by the medical establishment to be of direct relevance. Holistic approaches that integrate treatment are still rare. For children, the message behind this pattern may be that sexuality and questions of one's own sexual development are quite distinct from other areas of health care and physical development.

In this detailed division of practices, insurance coverage also transmits a message. What kind of complaints and illnesses are considered worthy of payment? What does it mean, for instance, that until a few years ago a mastectomy was covered by most insurance companies, but the plastic surgery following the operation was not covered? Although children themselves may not be aware of the exact distinctions made or the amount of costs, they see the effects of this consideration or hear discussions of it.

Another indication of the medical institution's values can be seen in the diagnostic process. What questions are asked, in what order, how much weight is given to each, and how the physician himself or herself reacts may all contribute to the patient's picture of the doctor's values and, in turn, to what is considered relevant to one's health. Are questions, for instance, about one's job, living arrangements, and daily routine considered pertinent to understanding a patient's health or diagnosing illness?

Finally, the values and standards of individual doctors, which they probably project to their clients, reflect both community attitudes and the upper-middle-class, often conservative, social status of their profession. These values, sometimes subconscious and deeply held, may be in conflict with or show an ignorance of the differing values of their patients.

Messages about Responsibility. Medicine typically focuses on individual illness, rather than on the physical or social environment. The profession thus strongly reinforces the notion of individual responsibility. Recent evidence on the role of the environment in producing and alleviating disease may change this focus somewhat. According to the National Cancer Institute, for instance, 70 to 90 percent of all cancers are environmentally associated.[16]

Responsibility for an individual's health, however, in traditional American medicine has been left almost entirely to the doctor. He or she is given—or procedures allow the doctor to take—responsibility for judgments on medication, when treatment should begin and end, who should administer care, and even how much the patient should know. Because the profession has been predominantly male, this particular paternalism has also meant that women's health and illness are described by men. In some areas of treatment, the result has been a history of misunderstanding and myth-making.

Children may transfer such an abdication of general health responsibility to matters of sexual health and understanding as well. Conditions directly affecting their well-being may appear to require a technical sophistication beyond their ability. The trend in self-help literature and the credence given to it, as well as the feminist encouragement for women to learn about their own bodies and take responsibility for their own health may help to change this tradition.

In terms of responsibility for children, the medical profession has generally tended to reinforce the primary responsibility of the mother. Advice books for parents written by doctors still illustrate this presumption, despite recent attempts to involve fathers more actively in child care.[17]

Messages about Body Image. Medical practice has probably been quite important in shaping general social attitudes toward the body. For many young children, the doctor, nurse, or health professional may have the important impact of showing respect for the body. The way in which medical personnel handle the child and their concentration on the child's well-being may convey the message that one's own health is of interest and is something of value. Such attention may also provide the child with information about the body that he or she may not otherwise receive.

On the other hand, American medicine has traditionally been oriented toward diagnosis and treatment of disease and toward drug therapy, rather than toward such areas as physical fitness and nutrition. Questions about sexuality, therefore, may only arise when specific care is required, in situations of pressing need, or when a particular drug is available—and not simply as part of the patient's own understanding of his or her health.

The medical establishment is also given authority for defining exactly what the health of our bodies really is and what constitutes normal functioning. As a result, the profession probably contributes to a general discomfort with our bodies, and to downplaying their creative, expressive potential. This has been especially true for women, whose pregnancies and childbirths have often been treated as sickness. New emphasis on preventative medicine and on natural approaches to childbirth may change some of these old attitudes.

Youth Organizations

Few data are available on the number of children who actually participate in activities of the major youth organization—Scouts, 4–H, Boys and Girls Clubs, CYO, and Little League. National headquarters may provide membership data, but because actual operations are completely decentralized in all these organizations, little can be said on a national basis about what a particular group does and what the members learn from it. National and regional organization headquarters are reluctant to discuss their role in sexual learning or even to admit that they have a role.[18] They are all extremely sensitive to the perogatives of parents and the norms of local communities. The organizations rely on volunteer labor and space often donated by churches. Their position at the national level, as stated in their nationally distributed publications, therefore, is to avoid the explicitly sexual and to express rather bland toleration for community values. No doubt considerable variation exists at the local level.

Even if the youth organizations do adhere to their announced goal of avoiding explicit sexual discussions, they can still be sending messages about sexual learning—about values and standards, gender roles, and images of the body. Such messages can be easily discerned in the handbooks and pamphlets of the Boy and Girl Scouts.[19] Because these organizations are so large and their publications so voluminous they will provide much of the material for this discussion.

Messages about Values and Standards. Explicit messages about the values and standards advocated by the Scouts are expressed in the Promises:

Boy Scouts: "On my honor I will do my best to do my duty to God and my country and obey the Scout Law; to help other people at all times; to keep myself physically strong, mentally awake, and morally straight."

Girl Scouts: "On my honor, I will try to serve God, my country and mankind, and to live by the Girl Scout Law."

and the Laws:

Boy Scouts: "A Scout is trustworthy, loyal, helpful, friendly, courteous, kind, obedient, cheerful, thrifty, brave, clean, reverent."

Girl Scouts: "I will do my best to be honest, to be fair, to help where I am needed, to be cheerful, to be friendly and considerate, to be a sister to every Girl Scout, to respect authority, to use resources wisely, to protect and improve the world around me, to show respect for myself and others through my words and actions."[20]

These traditional values are more liberal than traditional—emphasizing world citizenship, pluralism of religious commitment, toleration of others, concern for the environment and self-respect. The Girl Scouts have moved farther in the liberal direction than the Boy Scouts. Even the Boy Scouts, however, now interpret "morally straight" as "You are honest, clean in speech and actions, thoughtful of the rights of others and faithful to your religious beliefs."[21] "Reverence" includes respect for the beliefs of others, however different. "Loyalty" recognizes that "... loyalties conflict ... remember to look at both sides. Listen carefully to the arguments and then do what you believe to be right. Then you will be loyal to yourself."[22]

The promises and laws are not interpreted by either group as dealing with sexual morality. The Boy Scouts' physical fitness merit badge pamphlet does, however, contain one page on sex that is explicitly judgmental. It condemns masturbation as "not compatible with the Scout's attitude of reverence toward God and his own personal purity" and as a "flaw in character." Homosexuality is described as "not in keeping with moral cleanliness and the high ideals of Scouting."[23] Leaders are advised to investigate and discuss "incidents of sexual experimentation" in the troop, but to distinguish between "youthful acts of innocence, and the practices of a confirmed homosexual who may be using his Scouting association to make contact."[24]

The Boy Scout Handbook, in discussing "sexual maturity," advises boys to talk over questions about sexual matters with parents, doctors, or spiritual advisers. Leaders are advised not to take on sex education and to refer questions to parents or ministers. Nothing is said about the morality of heterosexual relations beyond that the boys may become interested in girls. The Girl Scouts seem to say nothing at all about any of these issues.

The absence of any recognition that boys and girls ask questions about values and standards regarding sexuality, of course, sends a message in itself—that is, sexuality and the issues that it raises are not appropriate topics for exploration and discussion by Boy and Girl Scouts. If this message is also communicated by local troops, it contributes to a climate where sexuality is denied and avoided. The legitimate concern about homosexual exploitation may also contribute to denial and avoidance of sensuality and expressions of affection, especially among boys and men.

Messages about Gender Roles. One set of messages about gender roles is transmitted by the sex composition of youth organizations.

Boy Scouts and Girl Scouts are obviously for boys and girls, respectively, even though some integration occurs at the older ages. Sports teams, such as Little League, have traditionally been for boys only, though a few girls' sports teams have always existed, and though a few girls are now making their way onto boys' teams. A few organizations, like 4–H, have always been open to both boys and girls; but the activities of the two sexes are often different.

The basic message is that boys and girls do different things, and do them separately. Boys and girls do not camp, or play sports, or have meetings together. The sex segregation of play and out-of-school organized activity may reinforce a sense that boys and girls are really very different in what they are and can become.

Youth organizations also send messages about gender roles through the activities they encourage. The Girl Scouts are more interesting in this regard, since three themes seem to be present: appreciation for womanly skills like cooking, sewing, and gardening; encouragement of femininity in appearance and manner; introduction of more progressive and egalitarian notions of gender roles. These now simultaneous themes may have developed sequentially during the course of Girl Scout history.

To progress through the ranks of Cadette (girls 12–14) scouting to First Class Scout, girls must meet the requirements of several "challenges" and merit badges. The first challenge, "social dependability," is classically feminine: it requires a "health and good looks plan," with attention to grooming, posture, and fashion; "know-how" for becoming a "gracious and competent hostess" and skills as a guest, including manners and the ability to make pleasant conversation. Other challenges, such as "emergency preparedness," "active citizenship," "community action," "environment," "international understanding," and "the out-of-doors" are gender neutral and similar to projects described by the Boy Scouts. Three challenges—"knowing myself," "my heritage," and "today's world"—explicitly invite exploration and consideration of different approaches to being a woman.

Among the merit badges, all three themes emerge. Badges are offered for the womanly skills of child care, food raising, and home nursing. Femininity would seem to be the theme of the good grooming, interior decoration, and social dancing badges. Opportunities to explore traditionally male areas are offered by aviation, handywoman, rock and mineral, and science badges.[25] Thus, Girl Scouts are exposed to a variety of opportunities to learn about both traditional and more egalitarian gender roles.

The Boy Scouts seem to be much more limited in their conception of appropriate gender roles. Strong emphasis is placed on the manly virtues and on traditionally male activities—competitive sports, active games, outdoor activities, crafts such as carpentry, and skills such as repairing machines. A merit badge in cooking, that emphasizes outdoor and camp cooking, is their only obvious concession to traditionally feminine skills. The "family living" skill award does include a child care component. Nonetheless, the illustrations in that section are sadly stereotyped—men painting houses and cutting grass, a woman washing dishes. The Boy Scouts do not invite their members to develop their nurturing or emotional selves, to consider such career options as nursing or child care, or to reflect upon the sexual division of labor.

Messages about Body Images. Scouting for both sexes has a long history of emphasis on physical fitness, sports, and outdoor activity. Cooperative and individual activities—camping, hiking, swimming, fitness—are stressed more than competitive sports, even by the Boy Scouts. While games (with individual or team winners and losers) are an important part of the program, leaders are encouraged to choose games in which everyone can participate and that give different scouts the chance to do well.

The Boy Scouts have long advocated individual fitness. They now require a hiking skill badge for Second Class Scout and a camping skill badge for First Class Scout. An Eagle Scout must include among his merit badges camping and personal fitness, swimming, or sports. The Boy Scout program seeks to develop healthy, competent bodies and to encourage activities that make boys, whatever their individual differences, feel good about their bodies. Inevitable competitiveness may hinder the achievement of this ideal among the less talented, but the articulated norms of the organization are to encourage every boy's skill and competence.

The Girl Scouts, too, do camping and outdoor activity. Personal fitness, sports, and camping, however, are not required for progression in the Cadette program. This may be because the Girl Scouts want to give explicit recognition to different talents and preferences, or to eliminate competition among girls. The result, however, may be that girls are provided fewer opportunities to appreciate their bodies, to develop fitness, or to learn sports.

The Social Service and Child Welfare System

A large proportion of children come in contact at some point with a school counselor; a substantial minority encounter at least a caseworker and, more likely, other service providers, because they receive public assistance. For a small minority of children the social welfare system has a dramatic effect on their lives. These children are defined as severely handicapped, retarded, disturbed, or delinquent, and children whose parents are defined as abusive or neglecting. They enter the domain of child welfare, mental health, and youth services departments, and the world of foster care and institutionalization.

National data on children in these situations are scanty. Estimates suggest that 250,000–300,000 children nationally were in foster care in the mid-70s. About 240,000 children in 1970 were in institutions of various sorts, including training schools and mental hospitals. About 1.3 million delinquency cases and about 151,000 dependency and neglect cases were brought in 1974.[26] Perhaps as many as half the delinquency charges were "status offender" or "persons in need of supervision" (PINS) complaints. These complaints—most often brought by parents, but also by teachers or welfare workers—can bring difficult but not criminal children into extended contact with the juvenile justice and social welfare systems. Delinquent and status offender children are most commonly placed on probation at home, but can be placed in foster care homes or group care.[27]

Although the number of children who encounter the child welfare system is small relative to the total number of children, the system's impact on their lives is obviously profound. In addition, the very existence of such a system sends messages to those outside it, through the widespread publicity often given to its activities.

Child welfare and youth services departments have historically had two purposes: to protect children from dangerous environments, and to protect the society from dangerous (or simply difficult) children. The two purposes have often been served by the same institution—the juvenile or family court—that had jurisdiction over both delinquent and dependent children. In most states today, however, the functions are separated; most typically, several state agencies take (or avoid) responsibility for the various types of children.[28]

All elements of the system have, because they deal with children, a protective and rehabilitative orientation. The system is staffed mostly by professional social workers who share values, standards, and practices as a result of common professional training. The system as a whole faces decisions about removing children from their families and placing them in foster care or institutional settings. It must distinguish appropriate from inappropriate homes, foster homes, and institutions. As a result, the system as a whole can be looked to for messages about values and standards, responsibility, gender roles, and life-style options.

Many of the attitudes that characterize the cash assistance system also characterize the child welfare system. (They are, after all, often part of the same state agency and are staffed by similarly trained professionals.) Ambivalence about standards, gender roles, and responsibility (described earlier) also characterize the child welfare system. In addition, however, the system delivers some uniquely powerful messages.

Messages about Values and Standards. The children in the child welfare system are there because they or their families are perceived as deviant in some way that is likely to be damaging to either the children or the society. In dealing with these children, caseworkers and courts are forced to evaluate settings to find those best able to alleviate the deviance. They have considerable discretion in defining good homes and institutions. Values and standards about sexuality enter into these evaluations. A parent's "immoral behavior" may, for example, tip the balance when a decision is being made to remove a child. Conventional sexuality seems to be a factor in evaluating foster homes.

Different expectations about male and female sexuality seem to underlie differences in the ways girls and boys are treated under status offender or PINS laws. Far more girls than boys are declared to be persons in need of supervision, and girls are often confined in institutions for longer periods of time. Complaints against girls are often those of promiscuity or sexual innuendo, which are never brought against boys. Fears about a daughter's sexual behavior may prompt parents to bring "stubborn child" or incorrigibility charges against girls. Provocative appearance or behavior is often enough to bring girls (especially poor and minority girls) into juvenile court, and to keep them under court supervision for longer than their "crimes" would seem to warrant.[29]

This punitive attitude toward some aspects of sexuality in girls may have several different effects. It may reinforce notions that sexuality is bad or dangerous. It may also serve to further separate the adolescent subculture from the conventional mainstream—that is, to strengthen the blatantly sexual aspects of the subculture and to increase the repression and intolerance of the adult mainstream.

Messages about Responsibility. The social work profession typically focuses on individual health and adjustment, rather than on the physical and social environment. The profession thus strongly emphasizes the notion of individual responsibility.

In the child welfare system, however, social workers' jobs are defined as caring for dependent, neglected, abused, and delinquent children. Workers are often confronted with poorly functioning and sometimes with dangerous families. The understandable tendency is to minimize the risks to children of remaining in their families by asking the state to assume responsibility. There have been many cases, both in the past and currently, where workers have too easily removed a child from the family. The current emphasis on due process rights for parents may change this somewhat.

Meanwhile, however, the child welfare system embodies a profound ambivalence about the responsibilities of parents and the state vis-à-vis children. It seems to be saying on the one hand, that parents are responsible for children, and on the other hand, that some parents are not competent to exercise their responsibility. These "incompetent" parents are disproportionately poor, minority, and single parents. The system thus sends a message that there are some kinds of people not fit to be parents.

Messages about Family and Life-Style Options. In making decisions about appropriate settings for children, child welfare workers inevitably make judgments and transmit messages about what they consider appropriate patterns of behavior. These judgments are subject to the whole set of conflicting pressures described in the first section of this chapter. On the whole, the system probably tends to reward traditional gender-role definitions and a conventionally middle-class life-style.[30] The liberal orientation of the profession, however, and the special problems of clients may make the child welfare system somewhat more progressive in these areas than the general norms of the communities in which the system operates. The insight that many nontraditional families are healthy and well-functioning has affected individual social worker attitudes and practice, and is now beginning to inform theory and training.

Because the decisions involved in removing children from families and placing them in foster homes are so important, the general tendency of the system to favor traditional middle-class patterns of behavior cannot be ignored. The attitudes of social workers are often in pronounced contrast to those of the subcultures from which their clients mostly come. These attitudes may serve to delegitimate options that would otherwise be open to the children of their clients.

IV. CONCLUSIONS

Few conclusions emerge clearly from this brief review of a few parts of the social welfare system. It does seem clear that the system embodies assumptions, attitudes, and values about important aspects of sexual development. The institutions and professionals in the system seem to transmit messages to their clients about values and standards, responsibility, gender roles, family and life-style options, and images of the body.

The messages, however, form no clear pattern. Some parts of the system seem bent on ignoring sexuality in all its aspects; some are inclined to reinforce what can only be called traditional sexual attitudes, gender roles, and life-style options; others are moving toward more progressive and egalitarian options. Some institutions appear to do all three, and most of the institutions considered here send mixed messages.

That the messages are mixed may well diffuse their impact. It may also mean that those who receive them, including children, perceive only the part of their content that is most compatible with already established attitudes and values. If this is the case, the social welfare system may act mostly to reinforce messages being sent by other parts of children's social worlds.

It is clear, however, that reforms in the social welfare system that make it more humane, more respectful of individuals, and more responsive to group differences would also result in clearer and healthier messages about values and standards, responsibility, gender roles, family and life-style options, and body images. Such reforms need not be advocated on the sole basis of their contribution to sexual learning; they make sense as good social policy generally.

The reform of cash assistance programs, for example, has long been on the national agenda. Welfare now stigmatizes and degrades the poor; it should be more adequate and more automatically responsive to need. Welfare and jobs programs continue to assume an unequal division of labor between men and women; they should encourage (or at least permit) women to hold productive jobs and men to care for children. Welfare regulations and welfare workers implicitly favor isolated traditional families; they should have more flexibility and appreciation of different family and household arrangements. The welfare system should encourage people, especially family members, to take responsibility for each other, without punishing the victims of irresponsible relatives. These principles will not be easy to build into a redesign of cash assistance programs. They ought, however, to be the goals of continuing reform efforts.

The health care system has come under increasing criticism for its attitudes toward women, toward sexuality, and toward individual versus professional responsibility. Measures for restructuring the profession—for changing its male-female composition, for emphasizing preventive care, and for altering client-professional relations—are likely to deliver more positive messages about sexuality as well as to improve health care generally. Reforms in the training of doctors to emphasize treating the whole person and to recognize explicitly the importance of sexuality could also improve sexual learning.

Youth organizations seem to be in a continual process of self-examination, moving to bring their practices in line with the current notions of sexuality and of male and female roles. They are not a target of government policy and should not be. Their efforts toward reform must continue to be generated internally, with the encouragement of their various constituencies.

The child welfare and foster care system is, as has been pointed out by several recent reports, often scandalous in its disregard of the well-being and rights of its clients. How the society treats abused, neglected, difficult, and delinquent children needs careful and humane rethinking. Here too, reform will be slow and difficult. As with cash assistance, however, reform of the child welfare system in accordance with principles that respect the basic humanity of parents and children will have the side benefit of contributing to healthier sexual learning for those who encounter it.

In short, the faults of the social welfare system's teaching about sexuality reflect more general problems in the system. In several areas, especially cash assistance, the problems are clear and the principles for reform understood. Those concerned about sexuality have an opportunity to support those working toward social reform for other reasons to bring about progressive change.

NOTES TO CHAPTER 7

1. U.S., Bureau of the Census, *Statistical Abstract of the United States* (Washington, D.C.: Government Printing Office, 1977), Table 497.
2. Office of Management and Budget, *1978 Catalog of Federal Domestic Assistance* (Washington, D.C.: Government Printing Office, 1978).
3. Census, *Statistical Abstract*, Table 70.
4. Ibid., Table 558.
5. Ibid., Tables 513, 543, and 191.
6. Ibid., Tables 497 and 498.
7. Ibid., Table 513.
8. Ibid., Table 543.
9. Ibid., Tables 194 and 196.

10. U.S., Department of Health, Education and Welfare, *Characteristics of State Plans for Aid to Families with Dependent Children* (Washington, D.C.: Government Printing Office, 1978).

11. Boston Women's Health Book Collective, *Our Bodies, Ourselves* (New York: Simon and Schuster, 1976).

12. Elizabeth Crocker, *Hospitals Are For Learning* (Halifax, Nova Scotia: Atlantic Institute, 1976), cited in Jonas Weisel, "Rx for Children's Hospital Fears," *Boston Magazine* September 1979.

13. Harold Lief, "Do All Medical Schools Now Offer a Thorough Sex Education?" *Medical Aspects of Human Sexuality* 13, no. 2, 1979.

14. Health Book Collective, *Our Bodies, Ourselves.*

15. E.J. Roberts, D. Kline, and J. Gagnon, *Family Life and Sexual Learning* (Cambridge, Mass.: Population Education, Inc., 1978).

16. Health Book Collective, *Our Bodies, Ourselves.*

17. This literature is well reviewed by Rhona and Robert Rapoport and Ziona Strelitz in *Fathers, Mothers, and Society* (New York: Basic Books, 1978).

18. Research Assistant Alicia Wheeler called the national headquarters of most of the major youth organizations and persisted on the phone until she reached someone willing to talk. No one admitted a role for their organization in sexual development, though most admitted that they were concerned with "character," "development," and so on. Most sent brochures and other literature which confirmed their general avoidance of the issue.

19. For the Girl Scouts, we looked at the Brownie and Junior Leaders Guide (both called *Worlds to Explore*, 1977) and the *Cadette Girl Scout Handbook* (1975 printing, 1963). (Girl Scouting is now divided into Brownies, age 6–8; Juniors, age 9–11; Cadettes, age 12–14; and Seniors, age 14–17.) For Boy Scouts, we looked at the *Scout Handbook* (eighth edition, 1977); Patricia Connally, et al., *Worlds to Explore: Handbook for Brownies and Junior Girl Scouts* (New York: Girl Scouts of America, 1977); *Cadette Girl Scout Handbook* (New York: Girl Scouts of America, 1975); *Boy Scout Handbook*, 8th ed. (New Brunswick, N.J.: Boy Scouts of America, 1977); *Scoutmaster's Handbook* (1972; reprinted 1978); "Personal Fitness" (1968; reprint 1974); and "Citizenship in the Community" (1972) in the merit badge series.

20. *Boy Scout Handbook* (North Brunswick, N.J.: Boy Scouts of America, 1977), pp. 34–51. *Cadette Girl Scout Handbook* (New York: Girl Scouts of the United States of America, 1975), pp. 10–14. If the wording of the Girl Scout promise and laws sounds unfamiliar to old Scouts, as it did to me it is because of a revision in 1972. The Boy Scouts retain ties to tradition.

21. *Boy Scout Handbook*, p. 37.

22. *Boy Scout Handbook*, p. 41.

23. "Personal Fitness," p. 12. Masturbation is described as follows: "Many young men at some time experience peculiar sensations from stimulating their penis. This is called masturbation."

24. *Scoutmaster's Handbook*, p. 74.

25. *Cadette Handbook.* It would be interesting to know how many girls actually earn each badge.

26. Census, *Statistical Abstract*; Maryke Allen and Jane Knitzer, *Children Without Homes* (Washington, D.C.: Children's Defense Fund, 1979).

27. An excellent description of the treatment of status offenders is given in Lis Harris, "A Reporter at Large (Supervised Children)," *The New Yorker*, 14 August 1978, pp. 55–89.

28. The complexity of the system, and the ways in which children are shuffled among various agencies, is described for one state in Children's Services Task Force *The Children's Puzzle* (Boston: Institute for Governmental Services, 1977).

29. Harris, "Reporter at Large"; and Barbara Brenzel, "Lancaster Industrial School for Girls," *Feminist Studies*, Fall 1975: 40–53.

30. An interesting study of attitudinal differences between caseworkers and their clients is Hope Lechter and William Mitchell, *Kinship and Casework*, (New York: Russell Sage Foundation, 1967). The study deals with attitudes toward family rather than toward gender roles explicitly, but many of the findings are relevant here, as well as to the area of family and life-style options.

 *Chapter 8*

Religion and the Sexual Learning of Children

*Sheila Collins**

INTRODUCTION

In attempting to assess the impact of contemporary religion on the sexual learning of children, it is necessary to say a few words about the origins and functions of organized religion in Western history.

More than any other social learning environment, organized religion—precisely because it exists to hand down a tradition—is the most impervious to change. Concepts, symbols, and images, as well as patterns of behavior, tend to get handed down from generation to generation long after the particular social context that gave rise to them has ceased to exist. In turn, traditional patterns of behavior tend to inform the way successive generations process their own reality. Thus, each Jewish synagogue contains an ark within which rest scrolls representative of the tablets given to Moses on Mt. Sinai, and Orthodox Jewish women, in spite of modern medical knowledge, are required to abstain from sexual intercourse for a period of time during and after their menses and to take a ritual bath before resuming such activity. For orthodox religious people of any faith conflicts may arise between the perception of reality taught by their religious tradition and the reality perceived by the contemporary secular world. On the other hand, in times of rapid social change, the more

*Sheila Collins is Administrator of the United Methodist Voluntary Service Program. As a teacher at the New York Theological Seminary, the Pacific School of Religion, and Union Theological Seminary, she has helped to develop feminist theory in the churches. She is the author of *A Different Heaven and Earth: A Feminist Perspective on Religion.*

orthodox traditions may also be viewed by some people as a refuge of certainty in a confusing and uncertain world.

Because religion attempts to explain when, why, and how all of human and nonhuman life came to be as it is, organized religion has, until recently, defined the boundaries of perception and cognition, legitimated patterns of human interaction, and set the limits of human conduct. Despite the fact that in the industrialized Western world religious institutions no longer command the overt civil power they did until the middle of the eighteenth century, religious mores and symbols continue to influence almost every level of our culture. Advertising images of women, for instance—sexual siren, virgin, or super-mom—are variations on the Christian themes of Eve the temptress and Mary the Virgin Mother. Laws that deny women access to credit, that assume women are the "natural" caretakers of children, that give men the family property in cases of divorce, or that discriminate against nontraditional erotic activity (such as sodomy) or nonprocreative intercourse are all derived from English common law, itself based upon ancient religious and legal proscriptions. Children who grow up in nonreligious or mixed-marriage families are still influenced by the dominant religious background of the culture.

More subtle than these influences is the often subconscious legacy of religious patterning. So deep is the religious current that socialist countries that have formally repudiated religion or countries that have embarked upon modern industrial development find religious prejudices interfering with their ability to create new social structures and a new human consciousness.[1]

The Kinsey studies revealed that nothing in the Anglo-American social structure has had a more profound impact on contemporary patterns of sexual behavior than its religious background. Generally, the impact of religion on sexuality has been to limit and channel erotic behavior. In the males and females studied by Kinsey, frequency of total sexual outlet was found to correlate negatively with religious orthodoxy and to correlate positively as the population became religiously inactive.[2]

Due to their nature and timing, the forms through which religion mediates sexual learning are likely to have a deeper impact on the human psyche than those mediated through almost any other learning environment. Western religion uses symbols, images, stories, space, gestures, and sound. In religious families, exposure to these forms begins in early infancy. Jewish boys, for example, are circumcised in a religious ceremony eight days after their birth. Christian children are often baptized in church before they can walk. Some of the first stories told to children are Bible stories. Though there is no way of knowing what particular effect prerational experiences have on the developing child, one must surmise that the general effects are deep and lasting.

Religious learning has a profound impact upon us not only because it makes use of prerational and nonrational mediating vehicles, but also because of religion's place in shaping the human response to those great imprinting experiences of the life cycle: birth, puberty, sexual intimacy, and death. Religion arose in the first place as humanity's attempt to make sense out of these sociophysical life crises. In such periods we are most acutely aware of the sensual dimension of experience as well as of the knowledge that through and beyond these rites of passage lies ineffable mystery. Inevitably, the language, images, smells, touch, and sounds through which we learn to interpret this experience will leave a memorable impression on us. No feat of logic later in life is likely to totally dislodge the influence of religious symbols from our subconscious—especially those symbols that mediate our sexuality.

In light of the tenacious hold that religious symbolism has on our culture, it is important to understand where the symbolism came from and why it arose. All three of the major religious traditions that I will be discussing in this chapter emerged from and have continued to be interpreted and handed down by male-dominated cultures. Moreover, they originated as a kind of reaction to a pagan polytheism that used overt sexual symbolism and invested all life—nonhuman as well as human—with numinous power. Androcentricism and the circumscription of "holiness" to a limited sphere are important concepts in understanding how and why Judaism, Roman Catholicism, and Protestantism have structured human sexuality as they have.

As modern feminists have repeatedly pointed out, androcentric religious traditions have perpetuated a false (because partial) view of human experience.[3] Not only have these traditions fostered dualistic and unequal gender-role identification (a dualism currently under attack in both popular and scientific circles),[4] but they have left unmentioned and unnamed those realms of sexual and religious experience that pertain uniquely to women—the experiences that arise from women's particular biology and social role. Childbirth, for example, which many women experience as a "holy" event, is seen nowhere in religious literature from the woman's point of view—except in certain Old Testament passages (long suppressed by androcentric religion) in which God is imagined either as a woman in travail or giving birth, or as a mother hen hovering over her brood. The male experience of sexuality, male-oriented language, and male roles have come to serve as the generic norm.

Though androcentrism has been a central feature of Judaism and Christianity from their very inception, not all religious traditions exhibit this lopsidedness to the same degree. Throughout their history, each of the three traditions that I will be examining has undergone permutations, so that within each tradition there are at least two major tendencies: the tendency toward orthodoxy and the tendency toward liberalism. The more orthodox branches of Judaism, Roman Catholicism, and Protestantism exhibit a more pronounced androcentrism than the more liberal branches, which have been influenced by changes in scientific knowledge regarding sexuality as well as by secular movements for women's equality.

Judaism is the fountainhead from which the attitudes toward sexuality that predominate in the Western world flow. Since it is the source of our understanding about sexuality, it is important to recall that Judaism arose from a patriarchal Middle Eastern nomadic culture and was modified as that culture developed a class structure and some cultural and geographic stability. The particular nature of the Judaic sexual code lies in its origins as a reaction to the Middle Eastern fertility cults. So far as we can ascertain, Middle Eastern fertility cults were the early human community's way of understanding its dependence on the cycles of nature for survival. In this world view, all human action was guided by the gods—usually symbolized in male and female form—who regulated the seasons and the cycles of human life and to whom appeal could be made for protection and guidance in carrying out the necessary functions of planting and harvesting, regulating labor, distributing the surplus, and conducting warfare. In many cases nature was anthropomorphized—the earth imagined as female, the sky as male, and the renewal of life in the spring as the union of male and female. Obvious physiological characteristics—especially genitals and breasts—were emphasized in the images and icons of the gods.

The emergence of monotheistic Judaism as a rationalistic, ethical religion meant that sexuality could no longer serve as an organizing symbol for many of the great life events. In the cosmology of Israel, the earth is not born from the womb of the primordial Mother, nor from the warring of male and female deities, as in other ancient cosmologies, but is created by fiat by a transcendent, rational God according to laws that can be discerned by humans and handed down from generation to generation. Nor can sexuality, in this world view, serve as the basis for interpreting the laws of nature. It is God who sends manna from heaven. So long as men (made in God's image) keep the covenant God has made with them, they will continued to sow and reap the harvest.

This is the beginning of a trend that distinguishes between the sacred and the secular. The uroboric female deity of the polytheistic world who symbolized the life-giving fertility of the earth and the life-threatening eruptions of nature has, with the advent of Judaism, been transformed into the purely human figure of Eve, who is shaped from Adam's rib by a God who generally comes across as masculine (though he is thought to be beyond all human sexuality).[5] In the Judaic world view, sexuality has been reduced from an interpretive schema to a volitional human activity meant to serve a utilitarian purpose.

The separation of human (profane) activity from sacred activity reaches its culmination in the early Christian church, which inherited from Greek philosophy and late Jewish apocalypticism a rigid sacred-profane or spirit-matter dualism that has influenced every facet of Christian life down to the present, particularly its ordering of sexuality. This dualism may be seen in separate and unequal male–female role patterns, in a fear of and disdain for the body and other manifestations of sensual reality, in rigid taboos on erotic behavior, in an elevation of mystical experience, and in some cases in a fascination with symbols of death. This dualism is more pronounced in Roman Catholicism, which is the direct descendant of the early Christian church, than it is in Protestantism, which today is largely a product of the Enlightenment.

This chapter will discuss the role of the religious traditions of Judaism, Roman Catholicism, and Protestantism in shaping the sexual learning of American children. To get the broadest picture of this role, four major subject areas will be examined. First, the source of religion's power lies in the distinct moral and ethical authority assumed by each tradition and the responsibility for decisionmaking that arises from that authority. What laws does a religion follow, and how do they guide the adherent in his or her sexual learning? A second area is the general message that a religion's teachings convey about human sexuality. How are erotic activity, masturbation, or virginity regarded? What rituals surround menstruation? What is said or not said about adultery and prostitution? Religion's description of appropriate gender roles is a third area of great importance. How do traditional rites and symbols help to establish separate roles for males and females both within the religion and in secular life? Finally, the very style of worship and the architecture of the religious center may communicate to children additional messages about sexuality.

AUTHORITY AND RESPONSIBILITY

Judaism

The constant break-up of Jewish culture and the dispersion of Jews throughout the world has given that tradition a particular flavor quite distinct from the Christian traditions that followed it.

Judaism incorporates the most comprehensive guide to life of any of the three religious traditions. A person's eating habits, sexuality, business ethics, social activities, entertainment, and artistic expression all fall within religious law and values.[6] The *Halacha*—the entire body of Jewish law encompassing both the Torah and the Talmud—contains laws and teachings on almost every conceivable aspect of human life. The regularity, rationality, and authority of these teachings has served to make the Jewish community a tightly knit subculture throughout its history of dispersion and persecution.

The extent to which Judaism has been resistant to modern, secular influences reflects the subcultural nature of the tradition and the desire to protect it from cultural extinction. As members of a particular Jewish subculture feel the need to guard their cultural and religious heritage from the influences of dominant Christianity or the secular industrialization that surround it, they may instill in their children an obligation to uphold the *Halacha* that carries the moral obligation to protect and preserve the community. Consequently, when an Orthodox Jewish boy breaks one of the commandments or *mitzvahs* required of him as a male, he is violating not simply an internalized rule, but a whole set of social ties, obligations, and responsibilities.

For Jews as a group, Orthodox and Liberal, the relationship between sexual mores and social responsibility is much stronger than it is in Christianity.

Roman Catholicism

Although Roman Catholicism shares with Judaism a dependence on tradition over individual interpretation and conscience, it differs from Judaism in giving even more religious and moral authority to its ecclesiastical leaders. Andrew Greeley, the Roman Catholic sociologist, has pointed out that within the Catholic community there is no social role for the independent lay activist such as those provided for the Jewish community by organizations like B'nai B'rith or the American Jewish Committee.[7] Moreover, within Catholicism there is no role for the layperson in the management, interpretation, or performance of social duties as there is in Judaism.

The dependence on clerical authority in Roman Catholicism tends to produce a sexual ethic that is more legalistic than that of either Judaism or Protestantism. Negative and positive behavior sanctions relate to specific physical activities rather than to intention or effect. Where the civil and religious authority of the church are combined, these sanctions will be more rigidly obeyed. However, where the civil authority of the Roman Catholic church is weak and the influence of secular culture pervasive, there may be a real split between the actual behavior of Roman Catholics and the articulated ethic of the magisterium. For example, Roman Catholics in the United States are known to practice birth control widely and even to contravene the Church's ban on abortion.

Protestantism

Protestantism places less stress on tradition and ecclesiastical leadership as sources of moral and ethical authority. The average Protestant is to be guided by a fourfold source of authority: scripture, tradition, reason, and conscience—or, as it is sometimes called, the "Holy Spirit." As it comes to us from the Enlightenment, Protestantism's ethical norms are to be internalized by the individual and adherence to them is generally decided by the individual, rather than by the community or an ecclesiastical authority. A child growing up within the Protestant community is therefore less certain about what his or her religion teaches about sexuality. This could lead to a greater openness toward new sexual and life-style options, or it could cause tremendous confusion and a moral vacuum that could lead to seeking more authoritative sources of decisionmaking.

There is a wide variation within the Protestant churches in the weight given to any of the four sources of authority. Anglican or Episcopalian children are likely to develop an ethical stance that is close to that of Catholics because much weight is given in those denominations to tradition and ecclesiastical authority. Unitarians, on the other hand, stress individual reason above everything else, while fundamentalist evangelicals place almost all the authority on a literal interpretation of the scriptures. Children who grow up in fundamentalist evangelical sects may internalize a set of strict sexual taboos similar in effect to those internalized by devout Catholics, but the source of those taboos is different.

Because of the diversity of authoritative moral sources within Protestantism and the wide range of practices that flow from this diversity, it is hard to talk about the effect of Protestantism "in general" on the sexual learning of children. We will, therefore, limit the examination of sexual messages found within Protestantism to those that apply to the mainstream denominations—Episcopalian, United Methodist, Presbyterian, Baptist, and Lutheran. By way of contrast, we will look at the Black church, which has a different history and ethos from the white-dominated, European traditions of the mainstream denominations. Class, cultural, and regional differences often distinguish the sexual messages conveyed by Protestant churches more than the particular denomination involved. This is why we can speak of a "Black church" even though Blacks are often members of the mainstream denominations.

GENERAL MESSAGES ABOUT HUMAN SEXUALITY

Judaism

Beginning with the second chapter of Genesis, "It is not good that man is alone; I will make him a helper like himself," the Hebrew scriptures emphasize the collective or relational nature of human existence. To be single is to be outside salvation. For Judaism, marriage is the paradigm of completeness; neither male nor female can be whole outside of this relationship. Judaism views erotic activity as one of the natural aspects of the human condition—an expression of the fragility and limitation that leads us to unite with others. However, if erotic activity is not carefully defined and regulated, instead of serving the function of social cohesion so important to Jewish culture, it opens up the danger of cultural dissolution. Therefore, erotic activity within legal marriage is blessed, but all forms of erotic expression which lie outside this relationship are viewed as "unnatural," "unclean," deviant, or dangerous.

While Judaism confines erotic behavior to legalized marriage, intercourse is seen to have more than the utilitarian function of providing children and thus passing on Jewish inheritance. The sexual act is viewed as a kind of religious ritual of commitment and solidarity between two people, mirroring the covenant between the people of Israel and God. It is an acknowledged attitude within Judaism that intercourse on the Sabbath is encouraged. In this light, sexual relations between husband and wife, if carried on within the proper restrictions, are viewed positively—as joyful and loving acts even apart from their relationship to procreation.

Prime responsibility for sexual activity within the marriage rests on the man as the one upon whom most of the Jewish ritual obligations fall. A whole set of instructions surrounds erotic conduct in Judaism. For example, men are forbidden to force their wives to submit against their will; they are proscribed from engaging in intercourse while under the influence of alcohol or if the couple is quarreling. Sexual relations are forbidden beginning with the onset of the wife's menstrual period and for seven days after its finish. A practice of having twin beds so as to avoid sexual temptation during the taboo periods has been common in Orthodox Jewish homes.

The elaborate regulations with which Judaism surrounds erotic sexual expression most certainly affect parental handling of a young child's exploration of his or her body. For example, since masturbation is severely tabooed by Judaism it is likely that a repressiveness exists within the Jewish family in regard to sexual play and polymorphous exploration. Such repression of childhood exploration will have an inhibitory effect on adult erotic behavior, so that in spite of Judaism's positive attitude toward sexuality within marriage it may be hard for adults who have grown up in a sexually repressive atmosphere to suddenly make the transition.

It is likely, too, that aversion to bodily functions will be stronger in girls than in boys. In simply avoiding masturbation, boys remain ritually clean; girls' natural functions—indeed, those associated with their main contribution to Jewish life—are associated with taboos of impurity. Nonorthodox Jewish women do not, of course, follow the prescribed ritual of entering a *mikvah* (ritual bath) to cleanse themselves of their impurity seven days after menstruation. Nevertheless, some subconscious residue of distaste toward these natural processes may be reinforced even in liberal or merely cultural Jewish women and men, simply because the proscriptions remain a part of the tradition. The authors of *The Jewish Woman in America* point out that "even today, Jewish women still live with the consequences of this patriarchal world view whether they have contact with traditional Jewish laws and customs, or identify themselves only as cultural Jews. For the conventional attitudes toward Jewish women rooted in these laws and practices are still transmitted from generation to generation through myth, behavior and unconscious responses."[8]

Some Jewish apologists, who have tried to point out the positive aspects of these bodily taboos, argue that because menstruation and childbirth are acknowledged by the tradition and accorded a ritual status, it is a sign that they are considered a normal part of human life; these events are not swept under the rug as they are in Christianity.[9] Kinsey and his coworkers, however, have argued that the tradition's openness about acknowledging sexual matters has little relation to the extent of sexual activity practiced by individual Jews. Orthodox Jews, in Kinsey's study, are the least likely to engage in erotic behavior of any of the groups.[10] Likewise, the fact that women's bodily functions are openly discussed may not necessarily mean that there is a less negative ethos associated with them.[11]

Finally, according to traditional Jewish attitudes, women's dress was to be modest and circumspect, exposing no parts of herself that are generally covered. For Eastern European Jews this injunction meant long sleeves, long dresses, and—because it was considered immoral for a man to look at a woman's hair—a wig to cover the hair, or to cover a shaved head.[12] A generally repressive attitude toward the body can be assumed from the dress of Orthodox Jews today. Although there is a distinct sexual difference between the requirements for male and female attire, it is not one that is meant to enhance the particular physiology of either sex, but to downplay the body as much as possible.

Roman Catholicism

In contrast to Judaism's positive regard for erotic activity, Roman Catholicism's fear of and disdain for sensual reality has elevated celibacy to the life-style most approximating holinesss, and has made marriage (the only form of heterosexual relationship condoned) a concession to human limitations. The church's dependence on the notion of apostolic succession and papal infallibility as its source of moral authority means that the messages it conveys both openly and subliminally regarding sexuality are developed and are enforced by a male celibate magesterium. Thus, a viewpoint on sexuality is conveyed by a group of men who have been taught to deny their own sexuality and who have usually been cut off from normal contact with peers during the crucial years of adolescence.

The message conveyed about erotic activity by the Roman Catholic church is generally a negative one. Erotic feelings partake of our baser nature. In the writings of the church fathers, which form the background to all Catholic teaching on the subject, eros is seen as a powerful demonic force associated with sin and death, that would overwhelm us and keep us from serving God if it were not held in check.

As a concession to humankind's limitations, the church teaches that the only legitimate purpose for erotic activity is procreation within marriage. Neither pleasure nor communion with another person are considered valid reasons for engaging in sexual intercourse. Erotic play among children, masturbation, premarital intercourse, extramarital intercourse, homosexuality, and any other kind of sexual activity that falls outside the strict parameters of the "legitimate" are considered sinful.

Because erotic conduct is an expression of the corrupt, mortal body, genital activity is not only associated with evil, but is also associated with death. Marina Warner, author of a book on the myth and cult of the Virgin Mary, asserts that updated formulas barely camouflage the ancient teaching that sex entered the world along with death with the disobedience of Adam and Eve.[13]

The association of sex with danger and evil is inevitably conveyed to Catholic children. The church's message is first beamed primarily at women, who make up the preponderance of worshippers in any parish. Women are assumed to be the sole or primary nurturers and teachers of young children—especially since woman as wife-and-mother or celibate nun are the only role models offered women in Roman Catholicism. Having been taught to fear their own bodies, Catholic mothers may convey in the home both subtle and overt messages that homoerotic play is taboo, that women do not have strong erotic feelings, that masturbation is a mortal sin, and so on.

The repression so early in life of natural curiosity about one's own body and those of others is likely to make it more difficult for the child to successfully transcend one of the crucial developmental stages described by Erik Erickson[14]—the stage of shame. Central to this dilemma is the church's contradictory assumption about human nature. On the one hand, it has taught that we are all made in the image of God and that children are innocent; on the other hand, we are innately sinful and unless strictly checked, will inevitably err in our ways. Children picking up such a mixed message are likely to grow up fearful of themselves and their own inner natures. For reasons that they are unable to comprehend, these children may learn that their own impulses, which they had assumed were good, are suddenly taboo and dangerous.

The Catholic church has long required Catholic parents, and even those of mixed marriages, to rear their children in the Catholic faith. Heavy pressure is also exerted to send children to parochial schools. Underlying such pressure is the fear of being tainted by the more "permissive" atmosphere of the public school, a fear rooted to some extent in the church's view of sexuality.

Some of the parochial school messages about sexuality are overt—such as those taught in hygiene classes. Eugene Bianchi, coauthor of *From Machismo to Mutuality*, recalls that he was taught to fear sexual impulses because they lead to loss of self-control.[15] Other messages are conveyed through exemplary role models—usually saints who were celibates, or saints who left their mundane family relationships to devote themselves to a higher "spiritual" calling. The example of an asexual Jesus who was born not of any mortal (sinful) union and an unstained "Virgin" Mary are the chief models the church has used to inculcate in its children the virtues of the good Christian. Messages about sexuality in parochial schools are also conveyed by the dress codes, the segregation of sexes, and by the use of celibates as teachers and administrators.

For girls, the option of a celibate life may have a special appeal. To commit oneself to virginity is to escape the corrupting influence of sex, while at the same time achieving a meaningful and independent role in the world—otherwise denied to those women who decide to "give in" to their erotic selves. Women who enter religious orders may do so in order to get away from the wearying and unloving sexuality that they have experienced in their own homes. In leading a workshop on changing gender roles with a group of nuns a few years ago, I discovered that almost all of them had grown up in large families where, as the oldest child, they had been required to take care of the younger siblings by a mother who was overworked and underappreciated.

Scientifically erroneous information about the body is apparently still being conveyed to children—especially girls—through the church's inculcation of the virtue of virginity. Marina Warner, for instance, describes symbolism associated with the virgin female body that presents the hymen as sealing off the womb entirely, thereby creating a pure and enclosed entity impervious to sin. In spite of modern medical knowledge, convent school girls may still be taught that their maidenheads are closed and the wearing of tampons will pierce the hymen and irreparably damage their virginity.[16]

Though virginity for males is encouraged, no such stigma attaches to the sexually active male body. Within the church's mythology, therefore, is the basis of a sexual double standard. Sexually active men have given in to their lust; sexually active women have become polluted.

Unlike Judaism's emphasis on positive *mitzvahs* and communal obligations, the Catholic ethical code is built upon a series of negative injunctions generally having to do with personal and interpersonal behavior. The rite of confession may be a way of ameliorating the harsh proscriptions of the faith. Here everything, including erotic activity, is forgivable if it is acknowledged as sinful and repented.

Because puritanical social systems often breed their own pornographic undergrounds, the underside of erotic repression, such as is found in the Catholic church, may be erotic obsession. Much of great Catholic mystical writing reveals a subliminated erotic obsession clothed in the language of spirituality, and modern clerical diatribes against women who are agitating for reproductive rights sometimes barely conceal a prurience about women's bodies disguised as moral righteousness.

Protestantism

In giving less weight to tradition, the Protestant ethos has adapted itself more easily to the values and dynamics (including those regarding sexuality) of Western secular and industrial capitalism than either of the other two religious traditions. In the case of Protestantism, we cannot as easily point to specific teachings, pronouncements, or symbols and their effect on sexual learning.

Though the Protestant churches often teach that the body is a temple of God, Protestantism shares with Roman Catholicism a general fear of bodily and sexual experience. The rigid body-spirit dualism of Roman Catholicism has been somewhat mitigated in Protestant experience largely because of the noncelibate nature of the ecclesiastical leadership. Within Protestantism, too, marriage is not seen as a concession to human limitation, but as the normative relationship for all adult men and women—and sexuality within marriage is considered normative human experience.

Nonetheless, in spite of the central part played by marriage and the family in the Protestant world view, this branch of Christianity also reveals a deep ambivalence about sexuality. This ambivalence is experienced in what is omitted by the religion rather than in what is specifically stated or legally proscribed. For instance, sex is rarely mentioned either in Sunday School literature for children or from the pulpit during Sunday services. Although sex education is expected to be conducted in the family, little guidance is offered to help the parents of young children deal with their children's growing bodily awareness and curiosity. Specific programs on human sexuality for junior and senior high school students have been developed by some denominations—those developed by the Unitarian and United Methodist Churches being the most comprehensive—but, in general, the needs of the younger child are ignored.

Where sexuality does enter the vocabulary of the young Protestant child in a religious setting, it may come through Bible stories learned in Sunday School. The Bible in fact is full of explicit references to sexuality, but much of its material—unless intelligently interpreted—is also likely to convey messages that are not helpful to a child living in late twentieth-century America. On the one hand, stories about the goodness of creation and the uniqueness of the individual before God are stressed in Protestant Sunday Schools. On the other hand, children may hear Bible stories that appear to blame women for prostitution and adultery, that stress women's subordination and obedience to men, or that place a stigma on women who are barren.

Since the social and historical context that gave rise to these stories and the viewpoint they reflect cannot be explained to young children (it is likely that the poorly trained Protestant lay teachers themselves do not have a grasp of this background), the pejorative references to sexuality and the denigration of women implicit in so much of the biblical material is likely to create feelings of mistrust and apprehension about sexual activity. "Sex is something that bad women do" is likely to be the concept conveyed through such stories.

Even the stories that are told to illustrate the forgiveness of God or the righteousness of a pariah may actually be conveying to Protestant children just the opposite kind of lesson. Consider, for example, the story of the adulterous woman as related in 8 John. The woman is brought before Jesus with the suggestion that her punishment be stoning. Jesus replies, "He who is within sin cast the first stone." The purpose of this Gospel story is to show how radically the ethic of Jesus differed from the legalistic moral framework of his contemporaries. This sophisticated understanding, however, is likely to be lost on a group of young children. Instead, the terrible sin of the woman—a breach of morality so severe that she could be stoned for it—is what is apt to come through.

GENDER-ROLE PATTERNS CONVEYED BY JUDAISM, ROMAN CATHOLICISM, AND PROTESTANTISM

Judaism

Children growing up within a traditional Jewish home are socialized into well-defined gender roles that carry with them responsibilities as well as rewards. More than the other two major religions, Judaism has preserved and created for each of the sexes rites of passage that ease and define the transition from one stage in the gender role to the next.

A child growing up in such an atmosphere is likely to be secure and emotionally stable so long as his or her inner inclinations conform to those gender roles established by the religious tradition, or so long as the religious tradition is able to maintain its hegemony over the lives of its adherents. Conversely, where Jewish children are subjected to influences outside the tradition, their ability to feel comfortable and unconflicted in those prescribed roles is eroded. For example, young people who have homosexual inclinations will find growing up within Jewish culture a deep source of conflict and inner turmoil. To admit homosexuality is both to admit something considered unnatural and to violate Jewish law.

For Jewish men, the norm of masculinity is generally associated with full participation in the public life of the community—participation in prayers and Torah study as well as business or a profession. The rituals of Judaic tradition that celebrate maleness, such as the rite of circumcision and the *bar mitzvah* may have a profound effect on the self-images and body orientation that both males and females develop. The fact that circumcision—a mark of full membership in the covenant—can only be performed on a male is bound to have a negative impact on female children. The joyous festivity surrounding circumcision must create in the female child a sense of inferiority, a feeling of lacking something that can never be made up through any effort of her own. For boys, on the other hand, taking part in this event and knowing that it was celebrated for them may provide both security in their masculinity as well as a feeling of superiority toward females who lack the vital requirements for admission to the male club. (The rite of circumcision, so precise in its acknowledgment of male genitality, may also be a key in the ability of Jewish males to withstand their ostracism from the male proving grounds of Anglo-Saxon culture. Jewish men may be more easily able to carry the traditionally feminine qualities of gentleness, passivity, prayerfulness, and studiousness because their masculinity was proven long before.)

The other imprinting event in Jewish tradition that celebrates the male is the *bar mitzvah*, or the marking of the boy's entry into manhood. The fact that this rite of passage is still celebrated gives the Jewish male a clear sense of his inclusion in the world of adult sexuality. Coming at a time when his body is going through major changes, this ceremony enables him to affirm and celebrate these changes and to connect them with the divine order of things and with the community into which he is welcomed. Since there is no comparable rite of passage in Christianity, we might assume that the *bar mitzvah* provides the adolescent Jewish male with a sense of security and importance, and a meaning structure that enables him to weather the storms of adolescence better than his Christian peers. On the other hand, the privilege during the ceremony of donning tefillin (reserved only for adult males to wear), and the reading of the Torah, may also give the adolescent boy a sense of superiority over even his own mother, to whom these obligations are denied.

The role of women within Judaism rests upon patriarchial privilege and the old-fashioned notion that biology is destiny. Since, until recently, women's lives were taken up with childbearing and rearing, women were exempted from most ritual and public obligations of Jewish law so as to be free to perform their presumed biological functions as wives and mothers. Within the private sphere of the home, women are required to carry out the dietary laws, the laws pertaining to ritual purity, and the lighting of the Sabbath candles. Their exemption from activities outside the family has meant that women have been excluded from defining the parameters of social reality.

Female children have no corresponding special ceremony marking their birth or entry into the community. Moreover, when girls are named, it is the father who—in the mother's absence—declares the daughter's name during a regular service in the synagogue.

Until recently, there was also no comparable, communal celebration of a female's entry into adolescence. To make up for this lack, Reform and Conservative Judaism have introduced for girls the *bas mitzvah*—a pale imitation of the male rite and an essentially meaningless ritual as long as women continue to be excluded from obligatory public religious responsibilities. *Bas mitzvah*ed girls are welcomed into no comparable "community." Moreover, unlike the boy who is called to read the Torah on the Sabbath morning, the girl performs her ritual on Friday evening in "a brief service in which the Torah is not even read."[17]

Historical evidence shows that Jewish women have not always accepted their inferior role within the Jewish community. Many middle- and upper-class Jewish women were enticed by the ideas of the Enlightenment to assimilate into the secular culture of nineteenth-century industrial capitalism, for that world appeared to offer them options that the Jewish community did not. Similarly, many poor and working-class women who joined radical secular or Jewish labor movements like the Bund did so because of the greater freedom and equality such movements offered women compared with the restrictive roles they were assigned in the Jewish ghettos of Europe.[18]

Only recently has official Judaism taken cognizance of the changing sex-role expectations in the Jewish community brought about by changes in the secular world. As a result of pressure from Jewish women, officials of the Conservative Movement—the largest branch of Judaism in the United States—announced in 1973 that women could be counted along with men in the minyan (the quorum required to hold religious services). At about the same time, the rabbinate was opened to women in the Reconstructionist and Reform traditions—the most liberal branches of Judaism. Conservative Judaism, however, still refuses to ordain women. The single female Orthodox rabbi in the United States was able to be ordained in defiance of tradition because, since it does not even consider such an ordination a possibility, the tradition has no legal statutes expressly forbidding it.

In spite of the legal gains made by women in the liberal branches of Judaism, women who aspire to nontraditional roles still must confront the weight of a tradition that opposes this course. The conflict between the desire to be a "modern woman" and faithfulness to the tradition has erupted in a Jewish feminist movement which, over the years, will likely make profound changes in the tradition's teachings on sexuality.

Roman Catholicism

Gender roles in Roman Catholicism cannot be understood apart from the three great symbolic figures—Jesus, the Virgin Mary, and Eve—that together embody all of the meanings the church attaches to male and female. Based on the church's inheritance of antiquity's spirit-body dualism, male was early on identified with "spirit," and female with "body." Because of the further association of the spirit with the godhead, males enjoy a privileged status in Roman Catholicism. Indeed, the prohibition against ordaining women is based on the "maleness" of Christ and on the belief that women are a lower order of being and cannot, therefore, represent God to the human community. The use of male language for God and the term "father" to refer simultaneously to God, the priest, and one's male parent serves to cement in the child's mind the superiority of men over women, and to reinforce patriarchal family roles.

Christ as the pure, unblemished Son of God is the primary male model for the Catholic boy. The images of Christ that predominate in Roman Catholic teaching and iconography are, however, either a sentimentalized infant or an agonized martyr—neither of which is likely to have much appeal for a boy growing up in a culture in which males are expected to be aggressive, competitive, and unemotional. It is likely, therefore, that few Catholic boys actually attempt to model themselves after the image of Christ.

The option of priesthood with its symbolic approximation to God does, however, offer Catholic males the chance to model other character traits—such as sensitivity, gentleness, and passivity—while retaining their position of male privilege.[19] Thus separated from the competitive secular culture, clerics can assume both traditionally masculine and traditionally feminine character traits without being considered abnormal.

Females, however, have a more limited range of characteristics and life-styles available to them. The sexual division of labor in the parish or parochial school—with women occupying clearly subordinate roles—carries a clear message to the Catholic youngster. Eugene Bianchi describes his own experience growing up in a Roman Catholic parish in California:

> The parochial Church and school taught me lessons, hidden and manifest, about the relationships between men and women. The division of roles in the parish was obvious. Priests, all men, of course, were of the highest importance and influence; they made the major decisions about parish policy. Even more significantly, they alone performed the saving functions of the sacraments. A ladies' altar society serviced the sanctuary and did other kitchen and cleaning jobs at parish functions. As youngsters, we automatically associated the priests with the realm of the mind, spirituality and public power. Women in the parish were examples to us of bodily concerns, lesser devotional piety and private, service roles. As a young boy, I could not have articulated the message that was taking shape in images and indirect symbols concerning women. In retrospect, the Church was preaching in a variety of ways its dual attitude towards women. The ruling male ecclesiastics both feared and deprecated women. Afraid of the ancient and mysterious power of the feminine, they treated her as a perpetual minor fated to be under the control of men and their institutions.[20]

The most powerful source of sex-role conditioning in Roman Catholicism is located in the myth of the Virgin Mary. The cult of Mary, symbol of ideal womanhood for both males and females, arose some time during the third century—probably as the response of the collective unconscious to the excessive denigration of the feminine in the ethos of early Christianity. By being both Virgin and Mother, Mary solves the dilemma of the mind-body dualism and the equation of the female with base materiality. In Catholic churches, convents, and schools, the figure of Mary appears everywhere as a desexed, medieval feminine figure with eyes downcast in humility or else looking down beatifically at the infant son on her lap. For Catholic males, taught to distrust the feminine in themselves and in women, Mary is a kind of compensation—a female for whom they can release their pent-up needs for emotionality, erotic experience, and dependency without having to deal with the responsibility and emotional complexity entailed in bringing these needs to real women.[21]

For Catholic women, the two sides of the Mary figure are impossible to reconcile. As Mother, she represents childbirth as the destiny of women. But as Virgin, she avoids the sexual intercourse that will obviously be necessary for any ordinary woman to have children. The effect, according to Marina Warner, is to create in women the need for the church's solace, and thus to further enlarge the institution's power.[22]

Mary's negative counterpart is Eve, whose seductive behavior is held up as a warning to Catholic girls. If, therefore, the only positive role model is that of Mary, devout Catholic girls are likely to grow up fearful of their changing bodies and of their sexuality—with a strong tendency to defer to male authority, especially the authority of clerics. Females who transgress sexual taboos—by engaging in premarital or extramarital sex, divorce, and abortion—are likely to feel that they are guilty of some deep sin and that no guilt is to be shared with their male partners. Women who are involved in working with rape victims and the victims of domestic violence report that Catholic women often exhibit a deeper sense of guilt over the violence inflicted on them than women from other backgrounds. In many cases, a strong Catholic background may keep a woman in a destructive marital relationship because she has been conditioned to believe that it is her fault if things do not work out.[23]

Through "First Communion," one of the first rites given to Catholic boys and girls at the age of seven, the church rehearses the bodily and emotional postures it wishes its adherents to assume later in life. For girls, the First Communion is a particularly strong imprinting experience, as it prepares a young girl for her future role as a subordinate and obedient "bride," either of Christ (as a nun) or of a substitute for Christ in the form of her husband. In many communities, girls taking their First Communion are dressed as young brides—complete with white dress, nosegay, and veil. The ceremony is surrounded with all the trappings of an actual marriage.

Protestantism

Like Judaism and Roman Catholicism, Protestantism tends (through its use of language, symbols, images, and church organization) to reinforce the patterns of the patriarchal family. Protestantism is probably closer to Judaism in conveying its message through an androcentric symbol system. The use of male language for the godhead implies a hierarchal order of being in which men are closer to God and superior to women. Such associations are further reinforced through the directive and sometimes authoritarian role played by the largely male clergy.

The positions that laypersons occupy in the Protestant church generally mirror the patriarchal patterns of secular culture. Men usually chair the church finance committee, the trustees committee, and other organs of churchwide decisionmaking and authority. As the ushers, they collect and count the Sunday offering. Occasionally they may sponsor churchwide functions or bring in outside speakers.

Women are typically Sunday School teachers. Women's groups or "circles" (which usually meet during the day, thus excluding any women who work outside the home) engage in church maintenance activities—cleaning church vestments, cooking dinners, and running bazaars. Often they also plan overseas mission projects or hold self-improvement programs. Although women's groups may actually keep the church together, their self-image is that of a service or auxiliary role.

The church appears to foster a kind of timidity in women that their role as homemakers, cut off from the competitive world of the workplace and the public arena, simply reinforces. Located physically in the same neighborhood as the home, cut off from the marketplace, the church can only see people in their private, personal, and familial roles. It does not recognize them as workers or actors in the public arena. The church, therefore, is a powerful reinforcer of industrial capitalism's separation of the home from the workplace and the divorce of ethics and morality from public policy.

Most adults who attend a Protestant church are assumed to be either married or widowed. A perennial complaint about the church is that there is no room for or recognition of people who do not fit into the typical middle-class patriarchal family role patterns. Thus, divorced people, single people, homosexuals, people living together outside of marriage, or very poor people are usually not seen in typical middle-class Protestant congregations. The message about who is included—whether articulated or not—is not lost on the churchgoing child. To belong to such a church family—to receive God's blessing through communal acceptance—means to conform to certain well-recognized behavior and life-style norms.

This message about the gender-role patterns conveyed by Protestant Christianity must be modified when we are talking about the Black Protestant church—especially Black churches of lower socioeconomic status. In Black churches generally, though the preacher occupies a central and patriarchal role, more give-and-take occurs between the largely female parishioners and the male leader. In a sense, the women accord the pastor a special position because they know the way in which racism has denied the average black man his dignity and masculinity. Since female members of Black churches are often the heads of their own families, or at least breadwinners along with their husbands, there can be little internalizing of an ethic of female timidity and subordination even though the same androcentric language may be used for the godhead.

THE EFFECT OF WORSHIP AND RELIGIOUS ARCHITECTURE ON SEXUAL DEVELOPMENT

Judaism

Judaism, with its acceptance and affirmation of material existence, differs quite markedly from Christian culture, which has at best tolerated materiality and at worst condemned it as the mark of sin and evil. For Jews, all of life—including the act of sex, the eating of food, the bathing of the body, the physical changes that occur at puberty, the harvest season—are occasions to recapitulate the community's history and to celebrate God's sustaining care. The practice of surrounding these acts or events with ritual and law gives them a positive communitarian meaning that they do not have in Christianity.

We may speculate that a child growing up in a family in which normally occurring physical events are accorded communal sanctity is more likely to experience sexual and aesthetic pleasure in connection with them than a child growing up in a culture in which the material world is viewed as inferior to that of the spirit or intellect. We may further hazard a guess that Jews (at least in the more liberal branches of Judaism) feel more "at home" in their surroundings, less conflicted about naturally occurring bodily changes and sexual feelings, more positive about intercourse with a mate, and less obsessed with sex than their Christian contemporaries.

The communal nature of Judaic worship is reinforced by the fact that most Reform and Conservative synagogues—in addition to being houses of prayer—are complete community centers that often include a swimming pool, a gymnasium, a kitchen, and various meeting rooms. The message conveyed is that religion has to do with the whole person—not just the head or the heart. It says, moreover, that taking care of the body is a religious obligation.

Nonetheless, despite the communal celebration of Judaic rituals and the communal use of synagogues, traditional gender roles are still preserved within Judaic worship and architecture. The long tradition of excluding women from the major celebrations in the synagogue—or, in the case of Orthodox congregations, secluding them behind a *mehizah* (partition) or in a separate gallery—has meant that women have taken little part in the chief areas of Jewish religious life. Because of the participatory nature of the Jewish religious service (the service may be and usually is conducted by laymen), Jewish boys and girls attending services witness their own fathers assuming the masculine roles assigned to them in the religious teachings. Seeing this patriarchal family relationship acted out week after week in a ritualistic tableau imbued with reverence and "holy awe" and with all the color and pageantry of a dramatic production acts as a conservative force on the aspirations of young Jewish boys and girls toward different gender roles.

The overwhelming sense of the synagogue as a male domain and the emphasis on "law" and "word" in Judaism did not, however, exclude a subliminal form of female imagery from the tradition. The Sabbath is often referred to as a Queen and as a Bride who descends to earth every Friday evening to fill homes with peace and love.[24] A friend who has recently been evaluating her Jewish heritage as part of an autobiographical novel has confessed that as a child she identified with the figure of the Sabbath Queen who comes to save her people. This symbol of a strong, positive, and active female force—coupled with her grandfather's assurance that she was "different" from other women—fed my friend's need for self-affirmation and fulfillment in the social world beyond the family, and enabled her to internalize only part of the restricted gender role offered by the tradition.[25]

Exploring the sexual imagery and symbolism so prevalent in Judaism, my friend also discovered that the ritual surrounding the reading of the Torah may be conveying more subliminal sexual messages. The Torah is a female symbol; indeed, it is often spoken of as "God's beloved, whom he made the bride of Moses."[26] During the service, the ark in which the "bride" has been enclosed is opened. The Torah is "undressed" of its female clothing—jewels, crowns, ribbons—and "made love to" by the male readers who bob over it with long pointers or *yads.* Though there is, of course, no consciousness on the part of the readers that they may be enacting ritual intercourse, the imagery and symbolism may be conveying just such a message to the unconscious. The subliminal message conveyed in this tableau may also reinforce the traditional view of erotic conduct in marriage—in which the women is a passive recipient of the loving, but dominant and more active, prerogatives of the man.

Roman Catholicism

Though there has been a great change in the mass since Vatican II (with such innovations as the use of the vernacular and the introduction of folk music), Catholic rituals continue to reflect the androcentric, hierarchial, and ascetic ethos of the tradition. Generally speaking, the mass is still a spectator event in which the priest—acting on behalf of the people—mediates between the laity and God. The effect of having the priest be mediator is to make the majority of the laity, who are neither clergy nor celibate, feel religiously inferior to those who have been set aside to do holy work. In the Catholic church a priest does not even need a community in which to perform his ritual tasks. No relationship needs to be formed between the people who attend mass, nor (unlike in Judaism) does a specific number need to attend in order to legitimate the rite. Before the introduction of English into the mass, the congregation did not even need to know what was occurring in the service. The separation of priest from laity and the superfluousness of community inculcates in the churchgoing Catholic the idea that holiness is attainable only through individual piety, and that sin is an individual act of commission or omission. The individualized yearning and guilt induced by such rituals is further reinforced in the confessional where one's individual trespasses are shared secretly with and only absolved by a priest.

The fact that all of the actors in a Catholic service are male and most of them celibate, reinforces the church's teaching that sexual expression has nothing to do with God and that women's erotic conduct, in particular, is a barrier to holiness. The use of a tasteless, paper-thin wafer to symbolize the body of Christ during the Eucharist (a wafer which until recently was placed in the open mouths of the communicants so that their hands would not defile it) inculcates in the Catholic worshipper the idea that material life belongs to our "lower" natures and that sensuality is somehow unclean, impure, and unholy.

Catholicism's lack of communal reference for holiness, sin, and guilt and its clear denunciation of aesthetic and sensual pleasure in both its teachings and its rituals have the effect of focusing the psychic energy of the average Catholic on his or her individual sexual activity. Guilt becomes unduly connected with individual acts that may violate the strict taboos of the church. Absent are the motivation behind the acts, the relationship in which they were situated, or any notion that sin may have a corporate source or that one's responsibility may be to the community.

The sensual pleasure and release that Catholicism denies, it reintroduces in sublimated form through a sensual preoccupation with symbols and images of death. The cross—with its mutilated body of Christ—is the central symbol in most Catholic churches, schools, and convents. Catholic architecture is meant to convey austerity, mystery, and awesomeness. Most Catholic buildings are large and cavernous. They do not lend themselves to intimacy or shared emotion. The usual posture in a Catholic church or convent is one of silence and bowed head. Though the pageantry of a Catholic service can be celebrative and quite beautiful—robed priests, incense, bells, candles, and choirs—the sensuality conveyed through such pageantry is ethereal, controlled, and muted.[27]

Children growing up in such an atmosphere soon learn that they must leave most of their emotions outside the church or parochial school classroom. Spontaneity, laughter, anger, an openness toward sensual experience, assertiveness and healthy self-regard are a kind of violation of the sanctity of most Roman Catholic space. The unarticulated message conveyed is that to be a good Catholic one must deny oneself the ordinary emotions and pleasures of life.

Protestantism

The setting in which most Protestant worship takes place speaks loudly of the religion's taboos. The worship center or sanctuary is usually a large, high-ceilinged room with rows of pews bolted to the floor and all facing the "head" of the church, which usually is occupied by the minister, the organist, and the choir. The typical mainstream Protestant service is directed by the clergyperson. The congregation stands, sits, or reads in unison at his or her cue. The parishioners have little chance to share emotion, spontaneity, or even genuine concern and compassion. With all eyes either facing forward or looking down at the program, the congregation is forced by the architectural setting to relate as individuals to the words being spoken or to the symbols used. As in Roman Catholicism, ethics becomes a matter of one's personal behavior, even though the words of the confessional prayer may begin with a "we."

Very often in Protestant services children are either excused from the service or are brought in for the first part of it and then dismissed for Sunday School. Protestants feel that children should not be forced to sit through an hour-long service that they may barely understand. The assumption is that children would get bored and fidgety, thus disrupting the solemnity of the service. Another message, however, may be conveyed to children. This message is that God dislikes noise, the free use of one's body, direct eye or body contact with others, spontaneity, joyfulness, anger (even when it is justified) or any of the real sorrows or triumphs that influence the daily lives of people. Moreover, God has nothing to do with what one's mommy or daddy does for a living, with bills that must be paid, with the wars that are mentioned on television, or with the conflicts and uncertainties of childhood. Though Christ himself said that unless you are like a little child you will not enter the Kingdom of Heaven, the child and his or her childishness is excluded from the adult worship of God. Only after the service is over and God has been dismissed do parents rejoin their children as people gather to face and greet one another—to act as a community rather than as atomized souls who face God independently.[28]

A frequent complaint about mainstream Protestantism is its lack of "friendliness," the lack of a feeling that a caring, supportive community exists. This complaint is probably most characteristic of northern middle- and upper-middle-class congregations, but less true of working-class and southern churches.[29] Since sexuality is rarely talked about in the average Protestant church, people who violate the never-articulated but subtly enforced taboos against premarital sex, extramarital sex, and homosexuality are likely not only to suffer deep guilt, but to have that guilt reinforced through the shame of silence or ostracism. Many of these people either drop out of the church entirely or seek other avenues of self-acceptance. While the mainstream Protestant church talks a good deal about "community," its silence on so many subjects that are the core of contemporary American experience conveys the message that there is little real community to be found here.

In contrast, the Black Protestant church, which grew up in the midst of the oppression of slavery, is a place to which people can bring what is most pressing to them and know that it will be received. Because Black churches are made up of people who are usually only marginally middle-class, and are generally located in the Black sections of cities, surrounded by all of the ills associated with poverty and racism, they tend to be more inclusive of the totality of human experience—including sexuality.

The hymns sung in Black churches generally convey a communal ethos rather than an individual one. The sufferings described in them are those of a whole people and the promises that are made are made to the race, not just to individuals. The sense of the church as a large extended family is conveyed through the use of the terms "brother" and "sister" to refer to other church members. This kind of language is rarely heard in the mainstream Protestant church.

Generally, the Black church is a place where emotions can be vented, whether of joy or sorrow. Frequently, the Black preacher plays on this need for emotional release through dramatic and histrionic delivery of the sermon. Gospel choirs raise and lower the emotional tone. The phenomenon of "falling out" appears to be a form of emotional catharsis through which a person who is deeply troubled can release his or her feelings in a communal and somewhat structured context. When this happens the person is supported physically by others and is eventually led by them back into a state of composure.

Children are generally not excluded as rigidly from the Black worship experience as they are in the white Protestant church. Since music is a big part of the service, there are often as many as two youth choirs participating. The strength and versatility of the music that comes from such choirs (compared with the often timid-sounding voices of children's choirs in white Protestant churches) says something about the dignity accorded to children in the Black church. In contrast to the white Protestant church, which often has its doors closed between Sundays (though now many churches house daily nursery or day-care centers, or are used by community groups), the Black church has usually been the center of the community's life.[30] Within the church's doors basketball teams are formed, classes in everything from reading to Bible history are taught, political forums are held, job counseling is done, and meals for the elderly are served. The minister is very often both a pastor and a community leader.

The young Black Christian grows up knowing that the church has historically been and still is one of the few authentic Black institutions in the community. Thus, the separation between sacred and secular and the split between individual and corporate ethics so pronounced in white Christianity do not hold for the Black church.

It is likely, therefore, that the mind-body dualism that underlies white Christianity is not operative in the Black church. Joseph Roberts, pastor of Atlanta's Ebenezer Baptist Church points out:

> The Black church has always been willing to admit that a person is not only cerebral but visceral. The White church has not. From the Greeks on, the White church has made a bifurcation between mind and body, so that everything that didn't fit into western, rational, logical categories could not be dealt with. So, how do you deal with a feeling? The White church just isn't able to deal with that. The psychiatrist is, and so you get psychiatry and religion as twin ministries. A person goes to a White church and stays for one hour on Sunday morning bleeding inside and then beats it to the analyst that afternoon or all the next week to try to get it together. The Black church always was able—sort of in the Hebraic concept—to keep mind and body together.[31]

From this analysis we may surmise that fear of the body, sexual guilt, and its corrollary, sexual obsession, are not as prevalent among Black Christians as among whites.

CONCLUSION

In a country characterized by high mobility, the increasing breakup of neighborhoods and community roots, a divorce rate of 50 percent, and growing economic insecurity, the religious institution may be the one remaining institution that is attempting to address the need for ethical direction, community, solace, emotional and physical security, nurturance, and healing. It is virtually the only institution set up to deal with the entire life cycle—from birth to death.

Certainly, these functions are what our religious institutions claim to be about. How capable are they, though, of meeting those needs? How capable are they of equipping young people to live in a society in which sexual violence is a routine occurrence on the nightly news, in which sex has become a commodity and is used as a gimmick to sell other commodities, in which more than half the women in the country must work outside the home in order to survive? How capable are religious institutions that sanction the patriarchal family as a model ordained by God of addressing the needs of the one out of seven children who will grow up in a single-parent family? When religious institutions allow the expression of only carefully controlled emotion and the admission of only ritualized and abstracted pain, how capable are they of healing the wounds caused by an increasingly precarious socioeconomic existence?

The interest of many young people in Eastern sects and cults, the popularity among members of the white middle-class of human potential groups, such as est, and the rise of an emotionally cathartic charismatic movement in Protestantism and Roman Catholicism indicate a failure of the mainstream religious institutions to meet the needs of growing numbers of people. One of the main attractions of religious sects, cults, and the human potential movement is the apparent community that they offer to people threatened with social atomization and isolation. In addition, by at least acknowledging the fact that their recruits may have been into drugs, sex, or other forms of "illicit" behavior, these movements may offer young converts the opportunity to resolve much of their confusion and guilt, even if resolving that guilt may mean adopting a rigid puritanism and cutting oneself off from one's roots.

In recent years religious leaders have begun to respond to the crisis confronting their institutions as a result of rapidly changing mores in the secular world. "Marriage Encounter" or "Marriage Enrichment" programs in both Protestant and Catholic churches have achieved considerable success in opening up communication between married partners. Greater communication and trust between parents is bound to have an effect on their children. Many seminaries now offer courses in human sexuality and a number of denominations have developed educational programs for lay people and clergy that use explicitly erotic films and other materials to "desensitize" the religious person who may be inhibited in talking about sexuality in a group.

Still, the extent to which homosexuality and women's liberation are hotly debated subjects within the religious community (the attention these subjects get far outweighs other issues such as disarmament, ecological disaster, or racism) reveals the continuing unease with which religion confronts the issues of human sexuality. In seeking to make sense out of human existence, our major religious institutions (perhaps with the exception of some elements of Judaism and Black Christianity) have, by and large, remained silently fearful about our capacity for eroticism. What poets and pagans have found full of holiness the Christian church has labeled "evil" and "dirty." In failing to provide a context in which sexual feelings can be acknowledged as legitimate, religion has repressed a good part of our capacity to love, grow, and create. Tom Driver, Professor at Union Theological Seminary, has pointed out that

> most of human sexuality (like other behaviors) is a learned capacity and needs cultivation by precept and experience. The church's greatest failure in this matter has been its refusal to nurture the sexual development of its members. The result is the totally irreligious sex that has plagued our history and is prevalent today.[32]

In some respects, the very wellsprings of the religious enterprise—religion's capacity to provide security and social cohesion by reference to a past source of authoritative "truth"—may prevent it from enabling people to adapt to changing realities in a way that is healthful. Religious ideology always lags behind changes in the scientific, social, economic, and political orders. Religion's reinforcement of ancient patriarchal role patterns—patterns that may have been functional for civilization when human existence was more precarious or when birth control was nonexistent—is a case in point. The legitimation of these role patterns by religious institutions now serves to fuel racial, sexual, and class divisions between people. Similarly, religion's sanctioning of Western male experience as the generic norm and its disregard for the traditionally "female" experience of nurturing lends weight to our civilization's headlong disregard for the ecological principles upon which human survival must be based.

Each of the religions that we have been discussing in this chapter began as a creative revolution against the dead weight of the past. Buried in each of the traditions are the seeds of that creative fervor. Many people who have grown up within these three traditions are now beginning—with women, who for so long have been denied the chance to define the parameters of religious and social reality, at the forefront—to recover those seeds and to reform those traditions in ways that, it is hoped, will enable religion to become once again a vital source of life and growth.

NOTES TO CHAPTER 8

1. Fatima Mernissi, "The Moslem World: Women Excluded from Development," in *Women and World Development*, eds. Irene Tinker and Michele Bo Bramsen (Washington, D.C.: Overseas Development Council, 1976), pp. 35–39; Teresa Orrega Figueroa, "A Critical Analysis of Latin American Programs to Integrate Women in Development," in *Women and World Development*, p. 46.

2. A.C. Kinsey, W.B. Pomeroy, and C.E. Martin, *Sexual Behavior in The Human Male* (Philadelphia: W.B. Saunders Co., 1948); A.C. Kinsey, W.B. Pomeroy, C.E. Martin and P.H. Gebhard, *Sexual Behavior in the Human Female* (Philadelphia: W.B. Saunders Co., 1953).

3. Mary Daly, *Beyond God the Father* (Boston: Beacon Press, 1973); Rosemary Ruether, *New Woman/New Earth* (New York: Seabury Press, 1975); Eugene Bianchi and Rosemary Ruether, *From Machismo to Mutuality*, (New York: Paulist Press, 1976); Sheila Collins, *A Different Heaven and Earth* (Valley Forge, Pa.: Judson Press, 1974).

4. Judith Bardwick, ed., *Readings on the Psychology of Women* (New York: Harper & Row, 1972); Jean Baker Miller, *Toward a New Psychology of Women* (Boston: Beacon Press, 1976); Dorothy Dinnerstein, *The Mermaid and the Minotaur* (New York: Harper & Row, 1976).

5. Several scholars of ancient mythology have traced the transformation of the female deity into the figure of Eve. See especially Raphael Patai, *The Hebrew Goddess* (New York: KTAV Publishing House, 1967); Theodor Reik, *The Creation of Woman* (New York: McGraw-Hill, 1960).

6. Hayim Habuy Donin, *To Be a Jew* (New York: Basic Books, 1972), p. 30.

7. Andrew Greeley, *The American Catholic: A Social Portrait* (New York: Basic Books, 1977), p. 7.

8. Charlotte Baum, Paula Hyman, and Sonya Michel, *The Jewish Woman in America* (New York: New American Library, 1976), p. 4.

9. Rachel Adler, "Tumah and Taharah: Ends and Beginnings," in *The Jewish Woman: New Perspectives*, ed. Elizabeth Koltun (New York: Schocken Books, 1976), pp. 63–68, is an example of this kind of apologist.

10. Alfred C. Kinsey et al., *Sexual Behavior in the Human Male* (Philadelphia: W.B. Saunders Co., 1948), pp. 483–486.

11. Baum, Hyman, and Michel, *Jewish Woman*, pp. 8–10.

12. Ibid.; Donin, *To Be a Jew*, p. 139.

13. Marina Warner, *Alone of All Her Sex: The Myth and the Cult of the Virgin Mary* (New York: Knopf, 1976), p. 151.

14. Erik Erikson, *Childhood and Society*, 2nd ed. (New York: W.W. Norton, 1963).

15. Bianchi and Ruether, *From Machismo to Mutuality*, p. 32.

16. Warner, *Alone of All Her Sex*, p. 74.

17. Baum, Hyman, and Michel, *Jewish Woman*, pp. 11–12.

18. Ibid.

19. Bianchi and Ruether, *From Machismo to Mutuality*, pp. 26–27.

20. Ibid., p. 27.

21. Ibid., pp. 30–31.

22. Warner, *Alone of All Her Sex*, pp. 336–338.

23. This observation comes from conversations with women I have worked with in crisis intervention centers.

24. Gail B. Shulman, "View from the Back of the Synagogue: Women in Judaism," in *Sexist Religion and Women in the Church*, ed. Alice L. Hageman (New York: Association Press, 1974), p. 147.

25. Conversations with Batya Weinbaum.

26. Patai, *The Hebrew Goddess*, p. 270.

27. This general description of Catholic ritual applies more to those churches that follow the northern European tradition. Mexican-American masses are often festive, colorful, and noisy, involving more lay participation.

28. Many Protestant churches have begun to incorporate a modern version of an ancient ritual called "the Passing of the Peace." This is a time in the worship service when members of the congregation are given permission to leave their seats, to mingle, shake hands, or embrace one another. However, it remains a fairly ritualized gesture, rarely allowing more than a momentary exchange between parties.

29. Horace Newcomb "Being Southern Baptist on the Northern Fringe," *Southern Exposure* 4, no. 3: 72.

30. When the white church does have its doors open between Sundays, it often rents the facilities to neighborhood groups that may not be integrated into the life of the church.

31. Joseph Roberts, "A Free Platform," *Southern Exposure*, 4, no. 3: 42.

32. Tom F. Driver, "A Stride Toward Sanity," *Christianity and Crisis*, 31 October 1977, p. 246.

Chapter 9

Human Sexuality: Messages in Public Environments

*Florence C. Ladd**

INTRODUCTION

This chapter focuses on environmental events and behavior that occur as sources of communication in everyday public settings. It is proposed that physical environments, both natural and designed, are sources of stimuli that contribute to various aspects of social learning, including learning about human sexuality. Such environments are omnipresent. This characteristic in itself suggests that the influence of environments may be appreciable. In this context, the term "public environments" refers to places and spaces to which anyone, without regard to sex or age, would have access. Streets, parks, museums, and public libraries are all examples of places termed public environments.

Several years ago, while on a trip through the People's Republic of China, I observed how dramatically public environmental messages convey broad cultural attitudes toward sexuality. The clothing of Chinese adults, for instance, revealed little about their sexual characteristics. The shapes of individuals were ill-defined under the formless shirts or jackets and trousers that both men and women wore.

*A former Assistant Dean for Academic Administration at the MIT School of Architecture and Planning, Florence Ladd has developed a special interest in the use of public spaces and designed environments. Dr. Ladd's articles have appeared in *Journal of American Institute of Planners, Journal of Environment and Behavior*, and *Landscape*. She has taught at Harvard Graduate School of Education and Harvard Graduate School of Design and is currently Dean of Student Affairs at Wellesley College.

Only a subtle bodily movement or perhaps, upon closer view, the hair style indicated whether the person was a man or a woman.

The clothing worn by children distinguished boys from girls somewhat more readily. Although both wore pants, the latter were dressed more colorfully and in styles and patterns that accentuated gender. During adolescence, many girls wear two long braids that are a reliable indicator of their sex, although their clothing at this age begins to become less distinctive.

In public interaction in China, either between adolescent boys and girls or between men and women, I saw little suggestion of sexuality. It was rare even to find an adolescent boy and girl together and extremely unusual to find them in a position that implied sexual attachment, that is, sitting close to each other on a park bench, embracing, or even holding hands. It was not at all uncommon, however, to observe the casual expression of affection between same-sex individuals—particularly among adolescents, who walked along streets or in gardens hand-in-hand, arm-in-arm or in a semi-embrace.

Public entertainment in the People's Republic of China provided slightly more evidence of sexuality. In theatrical productions, women and girls wore dresses or skirts that revealed their legs and arms; their songs and gestures, however, were like those of the male performers and expressed their affection for Chairman Mao, Chairman Hua, the workers, and the soldiers of the People's Liberation Army. There were scenarios that expressed devotion to family, brigade, or commune, but there was nothing personal in their content—nothing that expressed interest in another individual in erotic or affectionate terms.

Colorful political posters and billboards depicted people whose facial expressions and clothing reminded viewers of their patriotic duty. No commercial advertising appeared anywhere and, therefore, no need arose to use representations of people to sell products.

Western businessmen in Peking in the 1970s have reported that there are no red-light districts, no adult entertainment zones, and no evidence of or opportunity for illicit sex with the natives. Moreover, no female or male prostitutes cruise city streets in the new China.

Explicit expressions of sexual commerce were absent in the People's Republic of China. Implicit signs of erotic aspects of sexuality were infrequent, difficult for the outsider to decipher, and in general less evident than in other societies.

The contrast between contemporary China and the United States is striking. Drawing attention to the de-emphasis on signs and symbols related to sexuality in contemporary China sharpens the focus on evidence of sexual values and attitudes as well as interaction in public places in the United States. In comparison to the Chinese, people in the United States in the 1970s present a wide range of public behavior and signs that express an intention to communicate information about sexuality. The ideologies, styles, and conventions of the United States and the People's Republic of China are contrasting patterns that should be considered historically and culturally. Political changes and changes in fashion influence the way in which gender is expressed in "presentation of self" and in styles of dress. This suggests that the analysis that follows should be considered in terms of the historical and cultural factors that shape contemporary U.S. culture.

My observations in the People's Republic of China underscore the point that the cultural context is an essential framework for understanding the frequency and configuration of behavior, activities, displays, exhibits, and spatial uses that are related to human sexuality.

Anthropologists have identified patterns of events and behavior in a variety of societies in order to indicate the connections among patterns that determine the form and future of a culture. Foremost among anthropologists who have discussed the relevance of the physical environment to social learning and cultural forms is Edward Hall. In *The Silent Language* and *The Hidden Dimension* Hall's analyses of cultural differences in uses of space and interpersonal distance suggest the connections between behavior and environmental use.[1] Hall's works and R.I. Birdwhistell's earlier research in 1952 provided a basis for understanding the power of communication through body movements, postures, and interpersonal distance.[2] From birth onward, fundamental information about appropriate and inappropriate uses of space and body movements is subtly conveyed through human interaction. Thus, the physical environment, combined with the activities and behavior accommodated by that environment, contribute to social learning. In the broadest sense, the physical environment is educational or informative. Indeed, Amos Rapoport asserts that "the environment can be viewed as a form of non-verbal communication."[3]

Selected for analysis in this chapter are sources of messages in everyday environments that are significant in the transmission of cultural values regarding sexuality and sexual attitudes and practices. The twelve examples are varied, obviously quite distinct in their level of importance, and do not exhaust the areas that might be examined. Although most of the examples are drawn from contemporary U.S. society, many parallels are likely to be found in other Western countries. The assessment of the degree of the significance and consequences of exposure to such socially loaded elements in the everyday environment is beyond the scope of contemporary knowledge. At this stage, however, I wish to point the way toward their identification and eventually toward an assessment of their impact.

SOURCES OF MESSAGES IN THE PUBLIC ENVIRONMENT

Body Language

Not only does the rhythm of a person's gait suggest an element of sexuality, as several investigators of body language have observed,[4] but postures and gestures also convey a variety of messages—some of them with sexual content. Specific sexual messages are conveyed by how one sits or stands.[5]

Males and females have "vocabularies" of body movements that are distinctive. Although I know of no research that has determined at what age children begin to recognize and use sex-typed body language, there is no question that the recognition and enactment of bodily movements that are gender-related occur quite early in life. Consequently, an understanding of sexuality and sexual differences evolves from the recognition of bodily postures and gestures that are associated with males or females.

I would hypothesize that infants are handled differently by the same-sex parent from the way they are by the other-sex parent. When a father changes his daughter's diapers, his movements and points of contact are different from those when he changes his son's. Through their observations of where and how older children and adults of the same sex and of the other sex touch each other, where different forms of contact occur, and the affect associated with different episodes, children learn about appropriate and inappropriate expressions of sexuality. They begin to establish boundaries that limit the public and private expression of sexual feelings. They also learn about the everyday conventions that shape the interactions between same-sex and other-sex individuals. They develop an understanding of the meaning of physical distance between two individuals.

Hall describes distances maintained in encounters with others as follows: (1) intimate distance, (2) personal distance, (3) social distance, and (4) public distance.[6] Each of these categories also has a "close phase" and a "far phase." During the early years, probably in the first five years of life, children develop an understanding of the connotations of interactions between same-sex and other-sex people that occur at various distances. The recognition that affect, activities, and relationships are associated with physical distance is an important aspect of learning about sexuality. Through observing expressions of affection between individuals, noting, for example, that it is more common in the United States to see men holding hands with women and observing that boys and girls their own ages maintain "a certain distance" between each other, children come to understand the connections between sexuality and interaction.

Clothing

Even among young people—for whom "unisex shops" are a source of jeans, jackets, and shirts that emphasize male and female similarity—clothes are often clear signs of the sex of the wearer. In general, the styles, colors, and fabrics for clothing available in the United States (and throughout Western society) have been selected to emphasize sexual characteristics of the wearers. From the time pink is assigned to baby girls and blue to baby boys, it is clear that public acknowledgment of sex differences through one's manner of dress is appropriate, indeed expected. Moreover, settings and occasions in which sexual attractiveness is salient—such as at beaches and parties—clothing intended to heighten awareness of sex differences is the norm. Indeed an interest in being sexually attractive pervades U.S. culture. It appears to be a significant dimension of male-female interaction as well as female-female and male-male interaction in every aspect of daily life.

Clothing also plays a large part in conveying messages about sexual interests or intentions. Clothing is used for sexual emphasis in advertising to attract consumers. Scantily dressed female models (or male models dressed in clothing that suggests the rugged reserve associated with stereotypes of masculinity) stand next to cars, toothpaste, cigarettes, and beverages to draw attention to them and to suggest to consumers that these are "sexy" products and that you will be sexier if you buy them. To be sure, the objections from individuals and organizations supporting women's rights in recent years has increased awareness of the use of women as advertisement objects; yet women continue to appear in magazines and newspapers, on billboards, and on television in dress and situations that overstate the model's eroticism.

Clothing messages imply differences in attitudes toward male and female bodies. How much of the body may be left bare by the clothes and what bodily parts may be exposed vary with the season, the setting, and the sex of the individual. For example, while it is acceptable for boys and men to be shirtless when they are doing physical work, relaxing on beaches, swimming, or jogging, shirtless women and girls engaged in similar activities are termed "topless" and are not publicly tolerated. Children may appear nude at beaches, pools, and ponds until they reach the age of three or four (depending on the social context), when they themselves, their parents, or their neighbors decide that it is time to don bathing suits. Such subtle conventions are communicated at an early age.

Sports

In recent years in the United States, affirmative action awareness has made many parents and children conscious of the powerful message about human sexuality that is conveyed by participation in organized sports and other athletic events. In question now for many people is the stereotypic assumption that maleness is associated with greater strength and a greater capacity for athletic events than femaleness—an assumption that has been reinforced by the differences in atitudes toward the sports activities that are regarded as suitable for females and males.

With less separation of the sexes in physical education classes and some organized sports, the messages about sexuality generated by the "sports world" may be modified. The dominant message, however, will continue to be dispatched clearly: for example, that the athletic prowess of males is worth watching. Football, basketball, ice hockey, and baseball as played by men are likely to occupy the prime time of family life as well as prime time and space in the mass media for generations.

In coeducational schools and colleges, the prime playing fields will continue to be reserved for boys and men. Despite this different emphasis on sports in the lives of girls and boys, the increase in the participation of girls and women in events that demonstrate their physical capabilities should reduce the stereotyping of the relationship between gender and physical abilities. Although sports events with women participants may not claim equal time with men's events on television, the increase in exposure of women athletes will contribute to a change in the images of women.

Work Roles

Another form of information about human sexuality is derived from observation of the distribution of jobs in relation to sex. I am reminded of construction sites in the People's Republic of China where women and men shared the work of building houses and schools. There women share the "dirty work," but they seem to have no greater share in the visible "clean work" of their country than do women in the United States. Neither in the United States nor in the People's Republic of China nor elsewhere is there a significant number of women visibly engaged at a high level in public administration. The message regarding the relationship between sexuality and qualities related to high-level administration—such as decision-making ability, power, and political skill—is clearly male-oriented.

Except for their knowledge of their parents' work, children's contacts with people at work are largely with their preschool and primary school teachers, most of whom are women. It would be useful to explore the influence of women and men teachers on children's formulations of what is femaleness and maleness. More broadly, studies might examine the child's perception of the work world, given the present gender distribution. Studies are needed to determine the origins of concepts of the connection between jobs and gender. It would be interesting to learn what types of jobs boys and girls regard as desirable at an early age and to understand the extent to which the level of desirability is associated with the gender-appropriateness of the job. A study of jobs that are viewed as accessible or inaccessible to girls or boys because of their sex might be revealing. Such studies might take into account the sex-role of the teacher, and the difference between a woman teacher's influence and a man teacher's on the formation of ideas about work and gender. Certainly, there are many other sources of information that influence children's concepts in this area: textbooks, other mass media, values of peers, and their observation of the distribution of women and men at work in everyday life are all significant. More women than men are primary school teachers, receptionists, nurses, and waitresses in settings where children are likely to see people at work; while more men are principals, bus drivers, and doctors.

Analyses of the sexual distribution of publicly visible occupations that vary in status might yield interesting information about what is presented about sexuality through the observation of people at work. Studies of women and men in relation to work may expand our understanding of the characteristics associated with sexuality that are implied by the type of work that men and women do.

Political Demonstrations

The demonstrations for and against gay rights and abortion in recent years have brought into view people who publicly support positions on these issues. News coverage of the demonstrations has brought into U.S. family rooms and living rooms the terms "homosexual" and "gay"; and with them have come another dimension of sexuality that the children and adolescents of the 1970s and 1980s will understand more fully than their parents' generation. Some of those parents have probably found the search for words difficult when their children have asked "What's a homosexual?" or "Why are they called 'gay'?" or "What does 'straight' mean?" Slogans that summarize the purposes of demonstrations also introduce into the vocabularies of young viewers a collection of terms that refer to complex dimensions of sexuality. One wonders about their next level of questions and to whom those questions will be addressed. Sources of reliable information on homosexuality for children, parents, and teachers might contribute to a better and more uniform understanding about a part of society that is no longer closeted.

Museum Exhibits

Most science museums prepare exhibitions that intentionally convey information about human sexuality. The presentation of the male body and the female body is usually offered with straightforward clinical details, without a hint of mystery. Clean, well-lighted cases, show visitors the development of the human organism from a few moments after conception to its emergence through the birth canal. The accompanying text is factual and technical. The message here is candid and clear: understanding sexuality involves understanding complicated bodily processes and mastering complicated terms that seem to correspond to the simpler names that people already know for the parts of the body. Intentionally didactic, anatomical exhibits in museums are probably a useful contribution to the sex education of countless children.

But museums also contribute—unintentionally—a great deal to young viewers' images of sexual expression as interpreted by some of the world's great artists. A tour of sixteenth-to-nineteenth-century European paintings in a major art museum will include abundant representations of nude females—perhaps by Dürer, Titian, Tintoretto, Rubens, Courbet, and Manet. Women in these paintings are represented as objects or as victims by the men who painted them. John Berger has observed:

> in the art-form of the European nude the painters and spectator-owners were usually men and the persons treated as objects, usually women. This unequal relationship is so deeply embedded in our culture that it still structures the consciousness of many women. They do to themselves what men do to them. They survey, like men, their own femininity.[7]

Woman-as-object, therefore, is one message that the art of the human figure museum exhibit conveys.

Berger, among others, has noted that in Indian, Persian, African, and Pre-Columbian art, one finds the depiction of "active sexual love between two people, the woman as active as the man, the actions of each absorbing the other."[8] Thus, an understanding of the history of the treatment of the human figure by Western and non-Western artists, as well as the social and economic factors that bring forth the works of figuration art, should be a part of the essential education of every girl and boy in order to reduce the mystery and increase comprehension of representation of female and male figures.

Advertising

Advertising is another source of information available to children about how their bodies and their sexuality are regarded in their societies. Language symbols and human figures are used in ways that link sexuality to the material commodities and services that are advertised. This juxtaposition, combined with the suggestion that human sexuality is somehow relevant, may convey the message to children that sexuality is associated with the purchases and use of the objects or services advertised.

Billboards, posters, shop windows, radio, television, magazines, and newspapers bring into the everyday environments of children and adolescents a wide range of advertising messages. They are exposed to some messages that are not intended for them; nonetheless, they no doubt attend to and interpret in their own terms the images and messages that reach them. The content of advertising aimed at children is relatively free of sexual content when compared with advertising for adult markets. Still, it might be worth analyzing the sexual orientation of advertisements of foods, sports equipment, and toys presented through major promotional efforts.

We need to understand better the influences that determine the content of major advertising programs. The messages of major advertising companies reach large segments of the society. The conventions in the advertising industry regarding the content of advertised images and statements should be examined critically by panels of viewers. Interest groups, with viewpoints on sexual learning that should be taken into account, could make a valid contribution to the level of awareness and social responsibility in the advertising industry.

While the conventions of major advertisers can be monitored and perhaps influenced, it is difficult to get a purchase on the conventions among individual advertisers whose sexual requests may be published among the classified advertisements of newspapers and magazines. For example, "personal" columns in the classified advertising section, when deciphered by children and adolescents, may stir in them questions and personal searches that are premature. Studies of the readership of the "personal" listings might reveal interesting information about their audience and influence.

Adult Entertainment

When children begin to read and to ask questions about what they read, the hidden messages in adult entertainment signs may become puzzles to them. A cinema marquee announces X-rated films; a night club advertises "go-go girls" or merely "girls"; a shop offers "adult books." Situated near a suburban shopping center is a "massage parlor."

The dimension that renders these areas and materials "unsuitable" for children and youth is presumably their erotic content. Children gradually realize that because of their age an aspect of sexuality is off-limits to them. They probably recognize that "adults" pay for exposure to such material and that there is a slightly illicit quality to it.

Do they wonder or perhaps ask who are the adults who comprise the clientele for adult entertainment? Are Dad and Mom customers? Does the principal of their school seek adult entertainment? And just who produces and distributes adult entertainment? And why is such entertainment identified as "adult"?

The answers they receive or the explanations children themselves formulate regarding adult entertainment comprise another set of messages that contribute to interpretations of the treatment of sexuality in the society. One message is this: there is an illicit dimension of sexuality that lacks candor and clarity. Because it is illicit, it provokes their curiousity. They are probably curious about the values of an era in which the rights of women are being promoted in some quarters, while women in other quarters of the society are described as "go-go girls." The recognition of this paradox must provoke questions on the part of children and youth about the status of women in the adult entertainment business and the nature of the transactions between men and women in the context of adult entertainment.

The Newsstand

At your local newstand, at about the eye level of a six-year-old, adjacent to the display of superhero comic books are the most recent magazines from the pornography industry. The color photographs on the covers feature women and men with extraordinary physical endowments. They add an erotic quality to the otherwise prosaic environment of the newsstand. Their slick pages are filled with models and stories that go far beyond the pornographic literature available even a generation or two ago. More abundant, more varied, more sophisticated, and more attractive than ever before, the pornographic and near-pornographic literature currently available must be difficult for inquiring young eyes to resist. "Why does *Playboy* have a woman on the cover?" my six-year-old son once bellowed during a tour of a newsstand. His intonation italicized "boy." It was just the beginning of a series of questions and discoveries about sexuality to be made at our neighborhood newsstand.

No doubt pornographic literature has been the required reading in the informal sex education of children and adolescents since its initial appearance. While it is difficult to determine the content, quality, and quantity of pornographic literature available in the underground world of children, it is possible to determine what is available to them at the newsstand and to consider its message. The message is that sexuality has a glamorous, exotic, and illicit quality. The images are memorable and powerful.

Graffiti

The varieties of sexual interpretations that reach children through graffiti are lively, immediate, creative, expressive, and free. Graffiti with sexual connotations seem to appear relatively frequently in places accessible to children and adolescents, for example, in school yards, toilets, and so forth. Through such expression children learn that sex is worth writing about publicly and anonymously.

Initially, graffiti afford another medium through which they may practice reading, spelling, and drawing. Later some children and adolescents modify, respond to, and generate graffiti, some of which have sexual content. In some institutional settings expressly for adolescents or young adults—such as schools, teen centers, and college dormitories—washable boards and wall space are provided to accommodate graffiti. The institutionalization of graffiti, therefore, has legitimated graphic sexual expression.

Once limited to boys' and men's rooms, alleys, and school walls, graffiti in recent years have been written or drawn virtually everywhere. Now, on the walls of toilets for girls and women are to be found the graffiti of sexuality. In the bold color from spray paint cans, the sexual graffiti are writ large on sidewalks, bridges, rocks, and trees, on the walls of factories, housing developments, schools, shops, and subways. Sometimes elaborated in complex, esoteric statements, all of the graffiti of sexuality are not for all of the people. Still, there is a sufficient amount for the general reader with the skills and comprehension of a second-grader. A cross-cultural study of the sexual content of contemporary graffiti might reveal societal attitudes and norms about the anonymous public expression of sexual concepts.

Toilets

The separation and identification of toilets, labelled "Girls," "Women," or "Ladies" and those labeled "Boys," "Men," or "Gentlemen," implies that there is something unsafe, incompatible, or inappropriate about males and females using the same toilets in public places. The recognition on the part of children of separate toilet facilities in public places probably occurs at an early age. In restaurants, department stores, gas stations, libraries, and most schools separate toilet facilities are provided for males and females. While it is not uncommon for a mother to escort her young son to a room marked "Ladies" or "Women," it is unusual for a father to take his young daughter into a "men's room." To the very young, such a custom must appear to be a complex convention. The explanation that parents typically provide when called upon to clarify the convention probably only adds to the complexity.

Children know that at home family members and guests of the household of both sexes use the same toilets. The discerning five- or six-year-old who begins to distinguish between public and private toilet conventions must conclude that there is an element of sexual mistrust between females and males who are strangers. At a fundamental level, the information that they extract from their observations, from adult explanations regarding separate toilet facilities, and from adult warnings shapes attitudes toward males and females.

Implications

The examples selected here for analysis are only a few aspects of environmental uses that contribute to messages about human sexuality. A study of more depth might concentrate, for instance, on gender roles and examine within this topic territorial issues and the allocation of space. Such research would inevitably have to confront Nancy Henley's assertion that

> in general, females in our society have control over less territory, and less desirable territory, than males. . . . This is so . . . largely because of the privileges many men obtain by virtue of their status. . . . Males are likely to have higher status, and those spaces that come with it.[9]

The differences in occupational roles and relationships that are associated with sex underscore the territorial message. Henley observed that the images of a (male) boss dictating while the (female) secretary writes, or the (male) doctor operating while the (female) nurse assists him, or restaurants populated with waitresses serving men are also part of the everyday environments that shape concepts of sexuality and gender roles. Opportunities and environments for women, Henley notes, are "largely man-made and male-controlled."[10] It is not surprising to find girls and women are unintentionally and intentionally assigned to places and spaces that are less desirable than those occupied by males. The messages that underlie such systematic differences are relevant to the association of power, authority, and recognition in relation to sexuality. The mere recognition of systematic differences in territoriality and status associated with settings occupied by males and females communicates to young observers patterns of expectation about sex roles.

The postures and bodily movements of females with respect to males and vice versa carry messages relevant to an understanding of sexuality. An awareness of situations where females and males are separated communicates other messages. Environmental references to sexuality in verbal statements, uses of symbols, and representations of human figures convey yet a third set of messages about sexuality. The omnipresent character of such information raises questions about how all of it is interpreted, and what its potential impact is.

A general theoretical model is needed to account for the relationship between sources of environmental information and the distribution and impact of such information. With regard to environmental information with a sexual dimension, it is also important to account specifically for the mediation of such information according to differences in sex, age, ethnic background, and experience. Surveys, field studies, and laboratory research should be formulated to determine what environmental elements and events are perceived by children and adults. The perception of environmental elements and events with a sexual dimension should be analyzed for its impact on behavior and its contribution to the information children and adolescents acquire about gender and sex. In laboratory studies, films and simulated environments might include examples of art, advertising, and graffiti. Field studies in selected environments would allow investigators to determine specifically what environmental features with sexual content attract the attention of children and adolescents. Direct and indirect questions about environmental stimuli could be formulated for survey researchers who might interview children, parents, and teachers.

It is likely to be simpler to determine what children and adolescents perceive than it is to assess how their perceptions of environmental stimuli are interpreted. In early childhood, when most children are relatively open and direct, questions that require explanation and interpretation are addressed to parents, caretakers, and older siblings. The responses to the questions of early childhood direct the orientation of children and shape their future observations and questions. It is important to study the responses of adults and older children to inquiries about the everyday environment that have a sexual dimension. In this regard it is essential to understand fully what children and adolescents perceive, to learn about explanations offered them, and to study how they interpret what they observe.

When it is possible to assess the impact of environmental information relevant to sexual learning and development, we then will be faced with the difficulty of evaluating and using the findings. For example, how might we modify aspects of the physical environment to reduce, increase, or improve the content and quality of information about sexuality that is accessible to children and adolescents? What criteria could be used to determine what would be an improvement? What are the policy issues involved in determining how to modify references to human sexuality in public environments?

Although the impact of sexually relevant environments may never be assessed and managed in this society, it must be acknowledged that it is of consequence. The existence and recognition in public environments of sexually relevant material provides public validation of feelings, attitudes, and relationships that initially may be regarded as private, limited, or unique. Public displays of sexual information make evident the range of socially acceptable sexual interests, expressions, and behavior. Such presentations also suggest cultural standards and expectations regarding appropriate and ideal forms of sexual life. With respect to children and adolescents, information and evidence about human sexuality derived from public environments enlarge the range of legitimate concepts and behaviors that children recognize. The advice and information provided by parents, siblings, teachers, and peers may be validated or invalidated by a picture drawn on a subway wall or by an item read at the local newsstand or by the sight of lovers entwined on a grassy slope at the neighborhood park. The private sexual explorations and discoveries of children and adolescents gain credibility and a new significance when a representation of the experience is found in a painting, a photograph, or a cartoon. The messages extracted from a variety of public sources expand the context in which private experiences are interpreted.

The frequency of environmental messages pertaining to sexuality is determined by social conventions and political policies. The limited range of publicly explicit messages in the People's Republic of

China, in the Islamic world, and in other societies leads to speculation about the behavioral and attitudinal consequences of growing up in a society where public statements about sexuality are infrequent and rarely explicit. In the absence of public validation of sexual feelings and attitudes, presumably other sources provide information about the range and forms of acceptable sexual expression. Knowledge of both the sources and the content of messages with sexual themes that are conveyed to children and youth in socialist societies would contribute to our understanding of their sexual development and perhaps provide a reference for comparing developmental patterns to those of capitalist societies.

In general, cross-cultural studies of sources young children's sexual information would illuminate the factors that influence expressions of affection, childhood genital play, adult sexual behavior, aspirations for family life, and the variations on family life that occur in various cultures. To be sure, a thorough study of sexuality in our own culture would improve our understanding of sexual development, sexual behavior, and old and new forms of family life. An understanding of the influence of environmental messages would provide a context for the study of many aspects of sexuality in this culture and abroad.

NOTES TO CHAPTER 9

1. E.T. Hall, *The Silent Language* (Garden City, N.Y.: Doubleday and Co., 1959); E.T. Hall, *The Hidden Dimension* (Garden City, N.Y.: Doubleday and Co., 1966).

2. R.I. Birdwhistell, *Introduction to Kinesics* (Louisville, Ky.: University of Louisville Press, 1952).

3. A. Rapoport, *Human Aspects of Urban Form* (New York: Pergamon, 1977).

4. Albert E. Sheflen, Norman A. Ascraft, *Human Territories: How We Behave in Space-Time* (Englewood Cliffs, N.J.: Prentice-Hall, 1976).

5. N.M. Henley, *Body Politics: Power, Sex and Nonverbal Communication* (Englewood Cliffs, N.J.: Prentice-Hall, 1977).

6. Hall, *Hidden Dimension.*

7. J. Berger, *Ways of Seeing* (New York: Penguin Books, 1977).

8. Ibid., p. 53.

9. Henley, *Body Politics*, p. 36.

10. Ibid., p. 36.

 *Chapter 10*

Sexuality and Social Policy: The Unwritten Curriculum

Elizabeth J. Roberts

Our society today has no conscious, coherent, or comprehensive policy on educating individuals about sexuality. However, many of our nation's laws, statutes, organizational practices, community activities, and social mores are in fact based on beliefs and assumptions about sexuality. Such underlying assumptions might be best understood as constituting both our "de facto" national policy toward sexuality and the social curriculum through which sexual roles, relationships, behavior patterns, and lifestyles are learned in our society. For the most part, this "unwritten" sexual curriculum has gone unexamined, and public emphasis has focused on the role of the individual in achieving sexual understanding and well-being.

Left alone to make sense of complex and frequently contradictory messages about sexuality, many individuals of all ages have difficulty in understanding and managing their sexual lives. The emphasis on self-help and individual responsibility may only add to personal stress and anxiety. The need for a comprehensive examination of society's underlying assumptions and myths about sexuality can no longer be ignored.

We might like to believe that individuals achieve understanding, growth, and maturity through inner resources. However, this emphasis often fails to appreciate how dependent all of us are on the ability of society to create the conditions—social, economic, and political—in which fulfillment, responsibility, and satisfaction are possible.

The role of society in encouraging or limiting sexual learning becomes particularly conspicuous during times of social change, and today changes in the way Americans live their lives are clearly occurring. Personnel policies, divorce proceedings, rock lyrics, and clothing styles evidence changes in our sexual roles, relationships, and life-styles. These changes are not limited to the avant-garde academic community, to the eastern seaboard, or to the freewheeling California counterculture. For better or for worse, one is as likely to find evidence of changing sexual life-styles in small towns as in our fast-paced cities. The unmarried couple or the single parent, who may have once attracted attention for their nonconformity, are no longer considered exceptions, and can be readily found across the street or next door.

In light of these changes, many adults are confused about how to guide the sexual learning of young people. They have become uncertain about the accuracy of their information, the relevance of their beliefs, and the utility of the values they were raised to hold.[1] In a society in which marriage contracts are now subject to renegotiation or termination; in which changing definitions of masculinity and femininity are frequent mass media topics; in which male tears, tenderness, and compassion are slowly emerging as acceptable; and in which families may relocate geographically on the basis of the wife's work it is difficult to know what values or norms to pass on to young people to guide their decisions.

These difficulties are compounded by the fact that the outcomes of sexual conduct are less predictable today than in the past. Today, intercourse need not lead to pregnancy; premarital sex does not automatically ruin a woman's chances for a successful marriage; masturbation is no longer viewed as necessarily harmful to mental and physical development, but is accepted by many professionals as natural, healthy, and positive.

Most of us want today's young people to live personally satisfying and socially responsible sexual lives. In a world in which many of the taken-for-granted assumptions of daily life are now being questioned, and selecting from among a variety of life-style choices is becoming an increasingly large part of everyday living, we must help ourselves and our children become proficient and informed decisionmakers. Enjoying one's sexuality and taking responsibility for it requires decisions based on information and understanding. Unfortunately, the current conditions of sexual learning in our society have not encouraged such understanding.

In the past decade, three national commissions have stressed the importance of sexuality in one's life.[2] At the same time, the research studies prepared for these commissions reveal that mythology and misunderstanding about sexuality pervade our society, underlie our social institutions, and contribute to uninformed and misinformed sexual decisionmaking—particularly among the young.

For the most part, our official national response to this situation has taken three approaches: "disaster prevention" in the form of school-based sex education programs, "remedial treatment" offered by specialized community services, and "behavior control" largely achieved through religious and legal sanctions.

SCHOOL SEX EDUCATION

The role of the school in providing sexual information to young people was first considered over seventy-five years ago. Generally, battles over the form and content of such programs have waged intermittently ever since.

In recent years, many concerned national and community leaders have turned anew to the schools in seeking a response to the growing evidence of sexual ignorance and misunderstanding among children and adolescents. This continued focus on the schools is understandable. Their visibility, their public accountability, and the fact that they house our young for the better part of the day, five days a week, ten months a year for ten to twenty years make the school a logical environment for a planned educational curriculum.

Regrettably, this form of sexual education—when it is available—usually provides little more than factual information on "social hygiene" or reproduction. Sex-ed courses, designed primarily to prevent venereal disease and extramarital pregnancy, usually present endless facts about the function of the Fallopian tubes, the development of secondary sex characteristics, and the dangers of "social diseases." Rarely do such formal sex education courses encourage discussion of the many experiences, attitudes, and feelings that actually shape sexuality and give meaning to sexual experiences and decisions.

This narrow approach most likely derives from the view that the primary, if not only, purposes of providing young people with information about sexuality is to prevent the physical, social, and economic problems that can result from adolescent intercourse and pregnancy. "While these supressive aims are often unspoken, numerous contacts with the school and parent groups throughout the country indicate that they are often the real goals behind the drive for a sex education program."[3] Further, such programs seem to be based on the assumption that learning about sexuality occurs only between grades six and twelve, and involves only the acquisition of factual information. Although the concern for providing reproductive information to teenagers is understandable, it is misleading to suggest that the acquisition of a few facts on conception, birth, or VD can define the experience of learning about sexuality.

As suggested in the opening chapters of this book, many experiences and interactions throughout life influence values, beliefs, feelings, and decisions about sexuality. Furthermore, to understand fully, appreciate, and make sound judgments about the reproductive aspects of our sexuality, it is necessary to understand and appreciate other dimensions of our sexuality as well. For example, if a young person is taught that touching the body—especially the genitals—is shameful, all the factual information in the book may not help overcome the anxiety associated with using a condom, inserting a diaphragm, or having a checkup for venereal disease. Discomfort may be compounded if years of silence have persuaded that person that discussing sexual feelings is inappropriate.

Understanding sexuality involves understanding our feelings and beliefs about masculinity and femininity; it means coming to terms with how we want to express our affection or love for another person; it encompasses our notions about the erotic and our ease with emotional intimacy. Human sexuality is multidimensional, and sexual learning is a lifelong developmental process, not limited to formal, compartmentalized curricula.

Given the paucity of school programs that approach human sexuality in this comprehensive manner, it is not surprising that school-based sex education thus far has not resulted in either increased understanding for most people or even in the prevention of the social problems that it seeks most to eliminate.

SOCIAL SERVICES

Another common response to the current conditions of sexual learning is the development of community service programs designed to address personal and social problems associated with sexuality. In many communities, one can find programs to prevent venereal disease, counsel the victims of rape, deal with the consequences of unwanted pregnancy, or treat genital dysfunction. Although such after-the-fact, remedial programs make an important contribution to the community, they do not meet the full sexual learning needs of the population.

Since few alternative services exist in most communities, many individuals come to view sexuality principally as a source of problems. They may believe that seeking information, clarification, or guidance about sexual matters is only legitimate during times of crisis.

A recent study of service-providing resources in a large urban community concluded that most social services—regardless of their programmatic mandate—tended to equate sexuality with sex and reproduction.[4] Furthermore, service providers viewed these issues primarily as a source of personal, social, and economic problems. As a result, few individuals encountering these agencies have an opportunity to relate to human sexuality as a normal, enhancing, multidimensional aspect of life.

The study went on to analyze the setting of such services—that is, psychotherapeutic agencies, counseling centers, medical clinics, and hospitals—all places that are alien to the daily routine of most individuals. They concluded that such settings perpetuate the view that sexuality is set apart from the rest of life. Sexuality is not portrayed as a part of one's everyday roles, tasks, and activities, but is a specialized, isolated—and isolating—feature of the person. Additionally, the fact that so many programs are conducted through counseling and therapy facilities adds to the belief that if you seek information or guidance about sexual issues, "something is wrong with you."

Such a conclusion is further confirmed by the fact that most other service agencies in a community—such as youth organizations, recreation programs, neighborhood centers, or senior citizen programs—do not explicitly deal with sexuality at all. (This is not to suggest that these services may not be based on implicit assumptions about sexuality and communicate these messages in a variety of ways, but the lack of explicit programming by most organizations carries its own message.)

Because many services identified with sexuality deal primarily with reproduction, programs are usually designed for and used by women. Except for physicians, the great majority of people at the service level are women. This orientation towards women is reflected from program planning to the type of magazine likely to be found in the clinic waiting room. It makes little difference whether the service is related to counseling, birth control, or family planning or to pregnancy, delivery, and postpartum care. In all instances, reproductive services are designed for women. At best, men have a poorly defined role in this area of human sexuality. At worst, they are ostracized and made to feel like second-class citizens by the health and social service system. As a result, sexuality has become identified as a woman's concern, an issue that men do not need to deal with unless they become "victims" of a sexual problem.

This narrow identification of sexuality with reproduction or sex-related problems limits the development of more comprehensive and innovative programs that could help individuals understand the full range of their sexuality and improve the conditions in which sexual learning takes place.

RELIGIOUS AND LEGAL SANCTIONS

In addition to the "disaster prevention" focus found principally in school-based sex education curricula, and the "remedial treatment" offered by some community services, a third component of our society's explicit policy toward sexuality is its concern with "social control."

The two major institutions most frequently identified with controlling sexual life are religion and the law. Historically in our culture, most of the personal and social issues related to human sexuality have been understood as moral issues—that is, aspects of life that were to be interpreted, molded, and controlled through religious norms, sanctions and teachings. This history still informs our thinking about sexuality. Whether one currently subscribes to them or not, one must acknowledge that the moral codes and religious teachings based on the Judeo-Christian tradition are "far more than quaint relics of the past."[5] Religious teachings play an extremely important role in shaping our contemporary society's sexual attitudes and practices. To a large extent, social attitudes toward sexual pleasure, masturbation, premarital intercourse, homosexuality, contraception, divorce, family roles and responsibilities, abortion, and the role and status of women are guided and defined by religious precepts.

Earlier in this volume, the roots of many of these traditions have been traced to two fundamental aspects of early religious experience and teaching. The first aspect derives from the body-spirit dualism expressed in Greek philosophy and culture at the beginning of the Christian era. For the Greek, the good life implied an escape from the flesh into the superior life of the spirit. This dualism influenced early Christian understanding of the human self in ways that distorted the more integrated and holistic view reflected in earlier Middle Eastern cults. Consequently, dualism left an impact on Christian life and thought throughout the succeeding centuries.

The second aspect that helped to shape the Judeo-Christian tradition was the belief in the subordination of women. This belief was present in the institutions, the personal relationships, and the religious lives of patriarchal cultures. The body–spirit dualism found both cause and expression in this inequality. Because women were identified with emotion, body, and sensuality, the body-spirit dualism isolated women and associated their nature with a lower order of existence. While men retained undisputed control, whatever hinted of emotion, sensuality, and the life of the body became suspect, and something to be suppressed by those who aspired to a spiritual life.

The deeply rooted suspiciousness toward sexuality has lingered, and today we still derive many of our perspectives from this ancient mistrust of the sexual. The very notion of sin has a distinctly sexual suggestion to the popular mind. For example, the phrase, "living in sin," much more quickly suggests "improper sexual relationships" than exploitation of the powerless or the poor.

For most people, the power of religion over contemporary sexual attitudes and behavior is not expressed in the public arena, but within the individual conscience. Religion remains the great interpreter of sexual learning and expression through the unique role that its rituals and symbols play in profoundly shaping attitudes towards important imprinting experiences of the life cycle: birth, puberty, marriage, and death.

Religious teachings, however, have a social impact as well. Many public policy decisions about sexual issues are influenced by religious doctrine. For instance, campaigns mounted by various denominations to impact such current issues as the portrayal of sexuality on television, gay rights, abortion rights, or the equal rights amendment potentially affect religious adherents and nonadherents alike.

The public impact of religious teachings is also felt through the legal sanctions of our society. Based principally on a Judeo-Christian ethic, the laws and statutes (local, state, and federal) relating to human sexuality set the boundaries for what is approved and acceptable sexual conduct.

Most people probably are unaware of the many laws that define most erotic sexual experiences as crimes. Moreover, most people in our society are probably guilty of having broken one or more of these laws.[6] Consider the following data from the Kinsey study of male sexual behavior over thirty years ago: "At some time 85 percent of men had engaged in premarital intercourse, 59 percent had participated in oral-genital contacts, 70 percent had had intercourse with prostitutes, 30–45 percent had been involved in one or more extramarital affairs, 37 percent had participated in homosexual activities, an additional 10–15 percent had been exclusively homosexual for prolonged periods (3 years or more) and 17 percent of those raised on farms had had intercourse with animals."[7]

Kinsey's study of female sexual behavior, which was considered even more shocking at the time, shattered the notion of the asexuality of women. His study revealed that more than half of all American women had had intercourse before marriage, and that one out of every four American wives had participated in extramarital sex.[8] All of this behavior was (and is) considered criminal, and if rigorously prosecuted, could have resulted in the arrest of the majority of the population—a full generation before the mass media declared our society to be in a "sexual revolution." Patterns of private sexual behavior have not changed substantially since Kinsey first collected his data, and many policymakers and jurists have noted the numerous discrepancies that still exist between publicly expressed attitudes toward sexuality, the privately conducted sexual lives of most Americans, and the laws regulating sexuality.

Clearly a thorough examination is needed of our publicly accepted legal code regarding the erotic aspect of sexuality. However, attempts to introduce such review and change are often misinterpreted as promoting the conduct in question, and thus place reputations and careers in jeopardy—a risk few public officials are willing to take, particularly if they must run for elective office periodically.

Until such examination and re-evaluation occurs, the many criminal laws prohibiting specific erotic practices and defining the age, gender, and marital status of potential partners will remain on the books—subject to undesirable police practices and occasional capricious applications—thereby poorly serving the interests of society, the individual, and the law.

Each of the three approaches just described—formal school-based sex education, community services designed to treat problems associated with sex and reproduction, and moral and legal sanctions that control or regulate various dimensions of sexual life—comprise publicly acknowledged policies related to human sexuality. For the most part, however, these visible or formally recognized approaches have not increased understanding about sexuality (defined most broadly) or improved the conditions of sexual learning for most Americans.

In the first place, the narrow focus on sex and reproduction does not encourage individuals to understand the full dimensions and meaning of sexuality in their life. Rather, it encourages a view of sexuality as separate and isolated from most of human experience. It perpetuates the belief that sex (and hence our sexuality) is largely a product of inner drives—drives that without careful and constant control will result in myriad personal and social problems. For most young people, even before they reach adolescence, the underlying message of these policies becomes clear: sexuality is a negative and forbidden (albeit grown-up and exciting) dimension of personality, an aspect of humanity that can lead to guilt, anxiety, and emotional or interpersonal trouble.

Second, in general, these approaches only reach limited segments of the population. School-based sex education is usually available only in large urban and suburban centers. Where it does exist, for the most part it reaches only adolescents. Community agencies designed to address sexuality also do not reach the full spectrum of the population. The problem focus, the stigma attached to seeking help about sexual matters, the orientation toward reproduction and women, even the location of most services, keeps many individuals of all ages—particularly males—from using these resources. Religious and legal sanctions governing sexual conduct may have a powerful impact on personal conscience. However, the lack of legal distinction between consensual and coercive sexual encounters, as well as the obvious discrepancy between many legal and religious norms and actual patterns of contemporary life, may encourage young people to lose respect for the law in general or to disregard the moral dimension of human sexuality.

Finally, these policies regarding sexuality comprise only a part of our society's curriculum for sexual learning. In countless ways, they are supplemented—supported or contradicted—by other environmental influences. Television programs, newspaper columns, economic policies, medical practices, welfare regulations, youth activities, day-care facilities, advertisements, even the very structure of our physical environment—all play a crucial role in shaping an individual's sexual attitudes and behavior.

As social beings, we live among people within given social, cultural, and political structures. Our interactions with these people and systems can have an important impact on our sexuality. This is the process we have identified as sexual learning. A major premise of this book is that the programs, policies, and practices of our major social institutions are based on assumptions about sexuality, assumptions that are in turn communicated to today's young people. Institutions such as the school, youth and leisure organizations, or the media have many opportunities to influence childhood learning directly. In other instances, children may be influenced indirectly by the impact on their family of welfare programs, health policies, or work practices.

Each of Chapters 3 through 9 of this volume has examined a specific structural element of our society for its implicit and explicit contribution to sexual learning in childhood. The careful reader will have noticed that several distinct features characterize the messages communicated by this unwritten sexual curriculum.

In many cases, the assumptions about sexuality that underlie institutional practices and policies are not explicitly stated or immediately visible. In fact, they are unlikely to be recognized at the conscious level as related to sexuality. Nevertheless, these assumptions may underlie the choice of hours set for parent-teacher conferences or decisions about mandatory go-home-for-lunch programs at suburban schools. Assumptions about sexuality may be the basis for differing welfare or social security benefits for men and women or for the images of masculinity and femininity portrayed on billboards or in magazine advertisements. Because so many assumptions about sexuality go unacknowledged in our society, they can be extremely difficult to examine, evaluate, or change. As a result, it is easy to overlook the important role that such implicit messages play in the process of sexual learning.

Secondly, many of the assumptions that underlie our social institutions are based on a view of sexuality and social life that derives from Victorian times. As such, this view is not always relevant to the way in which many Americans live their lives today. "Respectable sexuality" in the late nineteenth and early twentieth century held that the family was central and women who did not save their bodies for marriage and children were "bad." For middle-class and even working-class women during Victorian times, a sexual misstep could result in social ostracism. At the same time, male sexual involvement with prostitutes or servants was tolerated, if not officially acknowledged. No respectable man, however, would trifle with a respectable woman. Legitimate sexual behavior was restricted to the marriage bed, with procreation—rather than pleasure—as its goal. Clearly, many (perhaps even most) Victorians did not live by these norms, and such strictures were even less characteristic of the attitude and behavior of the many immigrants from southern and eastern Europe who settled in America in the last century. Nevertheless, the Victorian ideology of sexual control and repression was the party line in America and came to be reflected in our laws and social institutions. This sexual mythology has been slow to change despite the growing evidence of social transition in many domains of sexual life.

A third feature of the conditions of sexual learning in our society is the inconsistent, and even contradictory, nature of many of the messages about sexuality. Certainly diversity, and even inconsistency, are inevitable and desirable attributes of a pluralistic society. However, when inconsistency is a symptom of antagonistic values or conflicting norms, it can create difficulties for individual learning. These difficulties can be further exacerbated when opportunities for discussion, clarification, and guidance are limited.

In light of this picture of the implicit assumptions, the current evidence of confusion and uncertainty should not be surprising. Sources of influence are not easy to identify. Sexual messages are not always clear and the values and attitudes communicated by one institution may be contradicted or minimized by the practices of another.

The essays in this book have attempted to initiate an inquiry into some of the more prominent of these institutional influences on childhood sexual learning. The data presented in these chapters help us to recognize some of the themes underlying society's sexual curriculum, and they also direct our attention to some of the potential problems inherent in our current approach to sexual learning.

Building upon this material, the final section of this chapter will underscore five specific areas in which current social messages about sexuality may limit personal understanding and add to the sexual malaise that is increasingly evident in our country. Taken collectively they suggest an agenda for future policy consideration.

GENDER SEGREGATION

Despite the popular debate and lively discussion carried on among professionals and in the mass media about the changing roles of men and women and the decline of the "double standard," most of the policies and practices of our major social organizations still are based on traditional and stereotyped notions of masculinity and femininity. Virtually every environment encountered by a boy or girl reinforces stereotyped assumptions about how men and women should feel, think, and behave. It is "remarkable how different the environment influencing the female is from that of the male from birth. These differences ever increase into an essentially dimorphic environment reflecting the sexual stereotypes."[9] From welfare practices[10] to magazine advertisements, from work requirements to Bible history classes, most children are taught that different rules govern the life-styles and conduct of men and women. In many cases, the emphasis on gender differences is so prominent that children come to believe not only that men and women are different, but that their interests, needs, and abilities are *opposite*.

Storybooks, toys, movies, and television identify femininity with passivity, nurturance, and emotional expressiveness. Female sexuality is presented within the context of love, marriage, and children. Erotic feelings and behavior are relatively underplayed in most female role definitions. Much popular lore still defines the "bad girl" as one who "gives it away, wastes her sexual capital, and ruins her chances to marry."

Masculinity on the other hand, is associated with a nature that is typically aggressive, nonexpressive, and wary of showing affection or sharing emotional intimacy. Males, supposedly uninterested in romance or family life, define their masculinity in terms of sexual conquests and scoring. "Interest in the desire for frequent erotic conquests—even if never realized—is part of the learned ideal for most males."[11]

In numerous subtle (and not so subtle) ways, children are warned of the problems inherent in demonstrating a trait or performing a role associated with the "opposite sex." The wife or mother who chooses, out of personal desire or economic necessity, to participate in the paid labor force is frequently the victim of inflexible working hours, the absence of quality day-care and an ethos that maintains it is inappropriate for men to assume responsibility for house and child care. The father who wants to participate more actively in parenting may find himself blocked by work schedules, pediatricians' office hours, and school practices. If, in the face of these barriers, he chooses to pursue his parent role, he may be needlessly forced to sacrifice a great deal in other areas of his life. Moreover, a male of any age who evidences nurturance, vulnerability, or sensitivity may be identified as unmanly. As a result, many children today may be learning to identify fatigue, stress, anxiety, or social ostracism as the necessary outcome of assuming nontraditional roles.

Children, observing these difficulties, may learn that opportunities for growth and personal happiness are necessarily limited by gender. The double standard associated with many of these traditional definitions of masculinity and femininity can hinder communication between the sexes and make it difficult for males and females to encounter each other as equals. Finally, children may learn that to seek a nontraditional role is not a discovery of joy and fulfillment but rather is a frustrating struggle and a test of will.

THE IDEALIZED STANDARD

Another set of messages embedded in our institutional policies and practices suggests to many children that there is one right way to organize one's life. From presidential directives that encourage unmarried White House staff members "living with someone" to marry, to the assumptions that underlie the design and construction of housing units, the "nuclear family" is the intended norm.

Because the uncoupled adult is viewed as an anomaly—perhaps occasionally even a threat to the social order—few provisions are made for single people (with or without children) in our society. The relative ease of marriage and the high cost of divorce, tax incentives and economic penalties, child welfare regulations, and municipal codes affecting communal and alternative living arrangements all discriminate against single people and alternative life-styles. Furthermore, being married and raising a family is often an unofficial job requirement for many jobs in public service or business.

Today, however, the "typical American family" with a husband-breadwinner, a homemaker-wife, and two children now makes up only 7 percent of the nation's families[12] and there is a growing tendency for young adults to either postpone or forego marriage. These changes, coupled with increasing incidence of separation or divorce, as well as the increased longevity of widow(er)s has greatly increased the number of single adults in our society. In fact, as of March 1978 the Bureau of the Census reports that one out of every five American households consists of just one person. This is a 42 percent increase since 1970. Census experts also estimate that the number of unmarried couples living together has doubled in the years between 1970 and 1978 and that nearly half of all children born today will spend a "meaningful" portion of their childhood with only one parent.[13]

And yet most of our major institutions are organized socially around the traditional nuclear unit. As a result, as the number of nonmarried persons, widows, widowers, divorced persons, and gay people increase, this organization of public life around the "normal" family unit and the couple may create even greater alienation for many Americans.[14] Children growing up in single-parent homes or in other nontraditional living arrangements may feel unnecessarily "different" or outcast for not living up to the idealized standard. And while children may learn that marriage fulfills important personal needs and social requirements, they may not learn to recognize the different costs and benefits inherent in every life-style choice.

SEXUALITY IN THE LIFE CYCLE

With our social concern with reproduction, it is not surprising that another theme underlying many of our social institutions is the belief that people are sexual only during a certain time of their life—their reproductive years. Baby care manuals generally neglect discussions about sexual learning and sexual development. Recreational, health, educational, and service organizations concerned with infants, toddlers, and children usually ignore sexuality and don't stress the importance of developing a positive sense of the body, ease with the genitals, and an accepting and trusting attitude toward physical and emotional intimacy. This denial of sexuality throughout childhood may add to the stress of adolescence in our society. Suddenly given the status of a sexual being, many teenagers are uncertain about how to manage this newly conferred status and for many young people, sexual behavior becomes a way to demonstrate independence, express anger towards parents and authority, or establish one's right to adult prerogatives. Sexuality becomes a proving ground and sexual expression a rite of passage.

At the other end of the age spectrum, the denial of sexuality can be even more acute. Newsstands, movie houses, and magazines all proclaim the relationship between youth, strength, beauty, and sexuality. As a result, many older people suffer negative attitudes toward their own sexual feelings and interests. Men and women who no longer measure up to some artificial definition of "sexy" are thought to have no interest in or desire for sexual relationships and activities. Physicians, social workers, or senior citizen programs seldom create opportunities for the discussion of the effects of aging or illness or the death of a partner on sexual roles, relationships, and behavior.

The problem can become most pressing when elderly persons are institutionalized and left without space or support for the expression of physical intimacy in hospitals and nursing homes. Important support is withdrawn at a time in life when there is important need for the intimacy that nourishes self-worth and the avoidance of isolation. As Alex Comfort so clearly states:

> Aging induces some changes in erotic performance but compared with age changes in other body systems, these changes are minimal. In fact, in the absence of disease, erotic capacity is lifelong, and even if and when actual intercourse fails through infirmity, the need for other aspects of the sexual relationship, such as closeness, sensuality and being valued persists.[15]

This denial of the relevance and importance of sexuality in childhood and old age is just one of the many opportunities for life affirmation that are denied the "powerless" in our society.

INADEQUATE COMMUNICATION

Reaching puberty or getting married does not magically guarantee that a person is knowledgeable about the "facts of life" or understands the meaning and consequences of erotic activity. Nevertheless, because childhood is alleged to be the "age of sexual innocence," few opportunities exist in this society for children to talk with adults about sexuality. Recent studies have demonstrated that parents today are no more likely than parents of a generation ago to discuss issues related to sexual intimacy with their child.[16] Moreover, similar silence pervades the child's encounters with pediatricians, Scout leaders, relatives, ministers, teachers, and counselors.

The few opportunities that children may have to hear what adults think or feel about erotic conduct or sexual intimacy are often limited to newspaper advice columns or community "hot line" telephone services. The anonymity built into such resources, however, may only reinforce the notion that sexual issues are not legitimate topics for inquiry or discussion, especially with those adults with whom the child is closest. Consequently, important dimensions of sexuality (such as masturbation, kissing, petting, intercourse, orgasm, wet dreams, menstruation, or sexual physiology) are left to the vagaries of peer communication.

Without accurate information and without the benefit of adult values, guidance, experience, and understanding, the actions of young people are apt to be error-prone. Satisfying expression of sexuality involves thoughtful decisionmaking. Much of the anxiety and distress caused by sexual behavior today results from lack of self-understanding and the inability to assess the consequences of one's behavior. While the child pays the price, the adult world sets the stage for these events.

Of course, some exceptions to this wall of silence can be found. Peers, as mentioned earlier, provide frequent opportunities for young people to discuss feelings and experiences and to share information about otherwise taboo topics. Although virtually no explicit sex is shown on television, there is considerable innuendo and verbal reference to the erotic dimension of life (often, however, in association with sex crimes!).[17] Movies, magazines, "adult" bookstores, and graffiti also are sources of sexual learning. Unfortunately, since these are often the only sources children may encounter, the overriding message may be the identification of the illicit with the exciting, the shameful with the titillating.

One conflict for children today is that they are deprived of sound sexual knowledge, while surrounded by sexual stimuli. This conflict, together with the general ignorance produced by our taboos on sexual communication, places young people at risk. The personal and social consequences of this situation can be felt by all of us —as we, our spouses, our friends, or our children relate to others who have not had the benefit of adequate knowledge and understanding about sexuality.

SEXUAL RESPONSIBILITY

Many parents, professionals, and policymakers today are concerned and distressed by the rising incidence of adolescent sexual activity. Often such concerns only thinly mask the deeper anger that many adults feel toward the irresponsibility that seems to characterize teenage sexual behavior. This irresponsibility is generally identified with lack of contraceptive use, rising rates of venereal disease, and premature marriages. Such anger, however, may fail to take into account the ambiguous, confusing, and contradictory messages that our society gives young people about sexuality and sexual responsibility.

Sports programs and athletic equipment, youth organizations, and adventure books encourage boys to initiate, to play aggressively, and to use their bodies to overcome obstacles. As boys get older, they find greater tolerance—and even approval—of their genital exploration, masturbation, or premarital sexual encounters. By adolescence, male expertise and knowledge regarding sexual matters is taken for granted. As a result, responsibility for initiating and guiding erotic conduct generally rests with the male; sexual success—measured in terms of both conquest and orgasm—becomes a measure of manhood.

The messages offered to young women are quite different. Billboards, department store window displays, movies, and the attitudes of parents and friends stress the importance of looking attractive and desirable. Genital play or masturbation among girls occurs less frequently than among most boys and is more likely to meet with adult disapproval. Female knowledge and understanding about erotic feelings and capacities is generally discouraged, and most females are raised to believe that "sex" is something that happens to them.

At the same time, many girls are warned about the dangers of the male "sex drive," and countless instructions make it clear that the female's responsibility is to control this "uncontrollable" drive and to prevent inappropriate sexual intimacy. Historically, this meant preventing any sexual contact outside of marriage. Today, however, the meaning of "appropriate" is less clear. It may mean no unprotected sex, no regular sex, or no intercourse without some demonstration of love or commitment to marriage. Whatever the message, the female is still generally required to enforce the operating norms or values of the community.

In the past, the male was responsible for initiating sexual intimacy and for protecting the couple from pregnancy. Although such protection did not always occur, there was a certain logic to this dual responsibility. Today, use of condoms is often replaced by the contraceptive pill or the diaphragm, and contraceptive technology now encourages the woman to assume responsibility for preventing pregnancy. Her responsibility in this matter, however, is complicated by the fact that she is not supposed to have erotic desires in the first place, much less plan ahead for their expression. As a result, a young woman who does engage in sexual activity today is faced with a dual dilemma: (1) she can acknowledge her erotic feelings, plan ahead for intercourse, and run the risk of being identified as promiscuous, or (2) she can live up to the still popular feminine norms of passivity and noninterest in sex, pretend she was "swept off her feet," and run the risk of pregnancy. For many young women, the source of the dilemma is not lack of contraceptive knowledge or contraceptive availability, but rather a definition of femininity that makes it difficult to acknowledge and plan for erotic feelings and behavior.

Stereotyped notions of masculinity and femininity take their toll on the young man as well. Although sex may be a male prerogative, his total responsibility for sexual pleasure can turn into a pressure to perform. When the asexuality of women was assumed, men had only to concern themselves with their own male pleasure. As female erotic interest and capacities have begun to be legitimated, however, many men have felt an added responsibility for their partner's sexual pleasure as well. Such pressure can lead to performance stress, anxiety, and even genital dysfunction and misunderstanding.

The impact of these antagonistic messages about pleasure and responsibility can be disastrous for both males and females. The male has been taught to initiate and to "score"; the female has been taught to resist. He has been urged to ignore her objection and overcome her resistance. They have both been led to believe that satisfying and meaningful sexual activity can occur without verbal communication and mutual decisionmaking. This mythology of the "battle of the sexes" is reflected on our movie billboards, in our religious parables, and even in our rape laws. Abortion clinics and the offices of sex therapists are filled with people who have been trapped by this socially sanctioned game.

In conclusion, our society assumes responsibility for the development of human potential through its concern for a physically healthy populace, an educated citizenry, and the work opportunities and civil rights of all people. If we believe that people who find satisfaction and self-worth in their sexuality are more capable than the sexually confused of dealing with life, and if we believe that sexual responsibility requires self-awareness and informed decisionmaking, then as a nation we must accept responsibility for the process and content of sexual learning.

This task is not an easy one. The confusion and difficulties that many people have in managing their sexual lives are the result of a complex set of circumstances and influences. Previous approaches to improving these social conditions of sexual learning have defined the issues and the audiences too narrowly to be meaningful or effective.

What is needed now is a fresh, concerted effort to place issues related to sexuality and sexual learning on our nation's agenda. Foremost among the priorities for action is the need to legitimate sexuality as an important, multidimensional, and ongoing aspect of one's humanity. The breadth of issues related to sexuality and sexual learning throughout the life cycle needs to be brought to the attention of the public in ways that will facilitate understanding and thoughtful action. Until we combat the misunderstanding and superstition that perpetuate the view that sex is a problem, we will not be able to increase understanding about sexuality for the majority of individuals.

Second, there is a need for policy change and program development based on the recognition that children need and have a right to information about all dimensions of sexuality, including the erotic dimension. For the most part, children today are exposed to a sexual curriculum that is based on gender stereotypes and ambiguous and conflicting attitudes towards pleasure and responsibility. Verbal communication is discouraged; discussion about erotic dimensions of sexuality is taboo. Such conditions are socially and personally irresponsible. Children are not given the information necessary to make important decisions or to manage their lives with satisfaction and responsibility. Parents, service providers, community planners, and policymakers must work together to break through the taboo of silence that surrounds sexuality, and to begin openly and honestly to communicate with each other and with our children about all aspects of sexuality.

Third, we need to recognize the implications of the changes that are occurring in the ways in which Americans organize and live their sexual lives. Too often institutional policies and practices serve to alienate, rather than support, individuals and families today. New economic incentives, work requirements, leisure organizations, and legal reforms are needed to provide support to the many individuals whose lives do not parallel the sexual myths of a century ago.

Finally, current efforts to examine the gender assumptions that underlie our social institutions must be continued and encouraged.

Too many of our young children are still learning to identify sexuality with a double standard and with limited, stereotyped definitions of masculinity and femininity. Parents who are attempting to expand their own role definitions or the roles of their children deserve the support of our schools, workplaces, and community services. The assumption that different rules should govern the behavior of males and females needs to be replaced with an attitude that emphasizes the importance of sexual equality, personal growth, interpersonal understanding, and mutual decisionmaking.

Human sexuality has had low status as an issue for serious rational public discussion for too long. The issues have been too easily dismissed as superficial, unimportant, embarrassing, or politically dangerous. Beneath this pattern of institutionalized hesitation, however, lies evidence of broad, if unfocused, public concern and need. Until those in a position to help focus public discussion and begin the needed inquiry become convinced of the importance of the issues, needed improvements will not occur.

It is obviously difficult to predict exactly what benefit changes in our social policies and organizational practices might bring to a new generation of children. In many cases change could provide opportunities for creativity and self-understanding that have simply not been known before. But if all the possibilities cannot yet be seen, the first steps toward them are clear. Such changes are first dependent upon the willingness of all of us to reassess our personal assumptions about sexuality and to increase our understanding of the current condition of sexual learning in our society.

NOTES TO CHAPTER 10

1. Elizabeth J. Roberts, David Kline, and John Gagnon, *Family Life and Sexual Learning* (Cambridge, Mass.: Population Education, Inc., 1978).
2. The Commission on Population Growth and the American Future, 1972; National Leadership Council on Venereal Disease, 1974; and The Presidential Commission on Obscenity and Pornography, 1970.
3. Esther D. Schulz and Sally R. Williams, *Family Life and Sex Education: Curriculum and Instruction* (New York: Harcourt Brace Jovanovich, 1979), cited in Judith Simpson, Lucille Aptekar-Litton, and Elizabeth Roberts, *Harmonizing Sexual Conventions: Service Providers and Sexual Learning* (Cleveland: Cleveland Program for Sexual Learning, Inc., 1978), p. 12.
4. Simpson, Aptekar-Litton, and Roberts, *Harmonizing Sexual Conventions.*
5. United Church of Christ, *Human Sexuality: A Preliminary Study* (New York: United Church Press, 1977).
6. Herant A. Katchadourian and Donald T. Lunde, *Fundamentals of Human Sexuality* (New York: Holt, Rinehart, and Winston, 1972), p. 421.

7. Alfred L. Kinsey, W.B. Pomeroy, and C.E. Martin, *Sexual Behavior in the Human Male* (Philadelphia: W.B. Saunders Co., 1948), cited in Katchadourian and Lunde, *Fundamentals of Human Sexuality*, p. 421.

8. Alfred Kinsey, W.B. Pomeroy, C.E. Martin, and P.H. Gebhard, *Sexual Behavior in the Human Female* (Philadelphia: W.B. Saunders Co., 1953).

9. Lillian G. Katz, "Guidelines for Teachers," in *The Sexual and Gender Development of Young Children: The Role of the Educator*, eds. Evelyn Oremland and Jerome Oremland (Cambridge, Mass.: Ballinger Publishing Co., 1977), pp. 298–299.

10. For a report on the ways in which the Social Security system could be adjusted to respond to the changing roles of men and women in today's society, see "Social Security and the Changing Roles of Men and Women," U.S., Department of Health, Education and Welfare, February 1979.

11. John Gagnon and Bruce Henderson, *Human Sexuality: An Age of Ambiguity*, Social Issues Series, no. 1 (Boston: Educational Associates, 1975).

12. "Typical Family Becomes Atypical," *Boston Globe*, 8 March 1977.

13. "Census Report: A Revolution in American Living Habits," *Sexuality Today* 2, no. 37 (1979), pp. 2–3.

14. United Church of Christ, *Human Sexuality*, p. 172.

15. Alex Comfort, "Sexuality and the Aging," *SIECUS Report*, vol. 4, no. 6, 1976, p. 1.

16. Roberts, Kline, and Gagnon, *Family Life and Sexual Learning*.

17. S.J. Franzblau, J.N. Sprafkin, and E.A. Rubinstein, *A Content Analysis of Physical Intimacy on Television* (Stony Brook, N.Y.: Brookdale International Institute, 1976); S.J. Franzblau, J.N. Sprafkin, and E.A. Rubinstein, "Sex on TV: A Content Analysis," *Journal of Communication* 27 no. 2 (1977): 164–170; and L.T. Silverman, J.N. Sprafkin, and E.A. Rubinstein, "Physical Contact and Sexual Behavior on Prime-Time TV," *Journal of Communication*, (1979): 33–43.

Index

About the Editor

Elizabeth J. Roberts, throughout her professional career, has been primarily concerned with issues in family life and public policy, with special interest in childhood learning. She has worked on the White House Conference on Children and Youth and has directed the Children's Television Unit of the Federal Communication Commission. As Executive Director of the Project on Human Sexual Development, committed to increasing the understanding of human sexuality of both children and adults, she has directed action/research programs concerned with the roles of parents, service providers and television in the sexual learning of young people. Elizabeth Roberts is President of Population Education, Inc. and now serves as Director of Television Audience Assessment, an innovative program assessing audience reactions to television programming. A well known speaker and lecturer, Ms. Roberts is the author of numerous articles on parent/child communication, sexual learning and the impact of television on children. Most recently, she has co-authored "TV's Sexual Lessons" and *Family Life and Sexual Learning.* The mother of an 11-year-old daughter, Ms. Roberts holds degrees in philosophy and education and is currently working on a book examining the conditions of sexual learning in the United States.